Valid Analytical
Procedures

Christopher Burgess
Burgess Consultancy, County Durham, UK

ROYAL SOCIETY OF CHEMISTRY

ISBN 0-85404-482-5

A catalogue record for this book is available from the British Library

Published by The Royal Society of Chemistry,
Thomas Graham House, Science Park, Milton Road,
Cambridge CB4 0WF, UK

For further information see our web site at www.rsc.org

Typeset by Paston PrePress Ltd, Beccles, Suffolk
Printed by Athenaeum Press Ltd, Gateshead, Tyne and Wear, UK

Valid Analytical Methods and Procedures

Preface

This handbook has been long in the making. Since the original decision to write it in 1995, much has changed in analytical science and progress made towards harmonisation of procedures and practices. However, the need remains for practising analytical chemists to adopt a formalism for analytical method development and validation embracing the necessary and sufficient statistical tools. The proactive role of the statistician/chemometrician in providing effective and efficient tools has long been recognised by the Analytical Methods Committee (AMC) of the Analytical Division of the Royal Society of Chemistry.

Analytical practitioners should be ever mindful of Sir R.A. Fisher's stricture that 'to call in the statistician after the experiment has been done may be no more than asking him to perform a post-mortem examination: he may be able to say what the experiment died of.'

As the title suggests, the intent is to provide a best practice approach which will meet the basic needs of the bench practitioner and at the same time provide links to more exacting and specialist publications. In this endeavour the author has enjoyed the support and active participation of the Chairmen and members of the AMC and of the Analytical Division throughout its long gestation. Particular thanks are due to past chairmen of AMC, Dr Roger Wood, Mr Colin Watson and Dr Neil Crosby for their enthusiasm and guidance. In addition, I am indebted to Dr Crosby for much of the material concerning the history of the AMC.

From the outset, Mr Ian Craig of Pedigree Petfoods Ltd and Dr Peter Brawn of Unilever Research, Colworth Laboratory have devoted considerable time and effort in scoping and shaping the handbook and, in particular, for generating and providing material on sampling and nomenclature. Without their unflagging support the project may well have foundered. On the statistical side, my thanks are due to Professor Jim Miller, President of the Analytical Division, who has been kind enough to read the manuscript thereby saving me from statistical errors and obscurities, and Professor Mike Thompson for providing the data set for the IUPAC collaborative trial example calculation. I thank Dr Dai Beavan of Kodak Ltd for allowing me access to some of their data sets for use as examples. Many other members of the AMC and the Analytical Division have kindly given me support and input including Professor Arnold Fogg, Professor Stan Greenfield, Dr Dianna Jones, Dr Bob McDowall, Dr Gerry Newman, Mr Braxton Reynolds, Dr Diana Simpson, Mr John Wilson and Mr Gareth Wright. I am grateful to Professor J.D.R. Thomas and Dr

David Westwood for their efforts in helping me ensure consistency and clarity within the handbook.

The help of Ms Nicola Best of LIC in Burlington House has been invaluable and my thanks are due also to Dr Bob Andrews and Dr Sue Askey of RSC publishing. My thanks are due to Paul Nash for producing the subject index. Finally, I wish to thank my wife and family for their forbearance during the preparation of this handbook and acknowledge financial support provided by The Analytical Methods Trust.

Contents

Valid Analytical Methods and Procedures

1 Introduction

1.1 Historical perspective

The development of standard methods of analysis has been a prime objective of the Analytical Division of the Royal Society of Chemistry and its precursors, the Society of Public Analysts and the Society for Analytical Chemistry, since the earliest of days and the results of this work have been recorded in the pages of *The Analyst* since its inception in 1876. An 'Analytical Investigation Scheme' was proposed by A. Chaston Chapman in 1902. This later evolved into the Standing Committee on Uniformity of Analytical Methods and was charged with developing standard chemicals and securing comparative analyses of these standard materials.

In 1935, the Committee was renamed the Analytical Methods Committee (AMC) but the main analytical work was carried out by sub-committees composed of analysts with specialised knowledge of the particular application area. The earliest topics selected for study were milk products, essential oils, soap and the determination of metals in food colourants. Later applications included the determination of fluorine, crude fibre, total solids in tomato products, trade effluents and trace elements, and vitamins in animal feeding stuffs. These later topics led to the publication of standard methods in a separate booklet. All standard and recommended methods were collated and published in a volume entitled *Bibliography of Standard, Tentative and Recommended or Recognised Methods of Analysis* in 1951. This bibliography was expanded to include full details of the method under the title *Official, Standardised and Recommended Methods of Analysis* in 1976 with a second edition in 1983 and a third edition in 1994.

The work of the AMC has continued largely unchanged over the years with new sub-committees being formed as required and existing ones being disbanded as their work was completed. In 1995, the Council of the Analytical Division set in place a strategic review of the AMC in view of the changing need for approved analytical methods and the need to develop future direction for the AMC as it moves into the next millennium.

The aim of the AMC was reaffirmed to be participation in national and international efforts to establish a comprehensive framework for the appropriate quality in chemical measurements, which is to be realised by achieving five objectives:

- The development, revision and promulgation of validated, standardised and official methods of analysis.
- The development and establishment of suitable performance criteria for methods and analytical instrumentation/systems.
- The use and development of appropriate statistical procedures.
- The identification and promulgation of best analytical practices including sampling, equipment, instrumentation and materials.
- The generation of validated compositional data of natural products for interpretative purposes.

1.2 Overview of the handbook

The objective for any analytical procedure is to enable consistent and reliable data of the appropriate quality to be generated by laboratories. Such procedures should be sufficiently well-defined and robust to ensure the best use of resources and to minimise the possibility of expensive large-scale collaborative trials yielding unsatisfactory results through lack of application of best practices. As part of achieving the objectives of the AMC it was felt that such a handbook would enable a consistency of approach to the work of the sub-committees.

Recently, major developments in statistical methods have been made particularly in the areas of collaborative studies and method validation and robustness testing. In addition, analytical method development and validation have assumed a new importance. However, this handbook is not intended to be a list of statistical procedures but rather a framework of approaches and an indication of where detailed statistical methods may be found. Whilst it is recognised that much of the information required is available in the scientific literature, it is scattered and not in a readily accessible format. In addition, many of the requirements are written in the language of the statistician and it was felt that a clear concise collation was needed which has been specifically written for the practising analytical chemist. This garnering of existing information is intended to provide an indication of current best practices in these areas. Where examples are given the intent is to illustrate important points of principle and best practice.

This handbook will be brief and pragmatic where possible. Inevitably, this will lead to contentious selections in parts. Consistency of a disciplined approach, however, is deemed more expedient than always espousing total scientific rigour.

1.3 Purpose and scope

The AMC identified the following four main objectives that this handbook should try to satisfy:

- Provision of a unified and disciplined framework that covers all aspects of the validation process from sample and method selection to full collaborative trial.

- Compilation of a selected bibliography of more detailed and specialist works to be used when appropriate and incorporating the work of the Statistical Sub-committee.
- Guidance in the use of the selected statistical procedures for the comparison of methods where circumstances and resources do not permit the meeting of the requirements of the IUPAC protocol.
- Illustration, by way of worked examples, of the main statistical procedures for the calculation, display and reporting of the results.

Analytical chemists are by nature innovators and seekers of improvement. In the development area these qualities are invaluable in optimising method performance. Alas far too often, this desire for continuous improvement spills over into the interpretation of methods for quality control. Here we require consistency of application and rigorous control of processes and procedures. These aspects are anathema for many practitioners of the 'art of chemical analysis'.

Whilst this may be sustainable (albeit undesirable) for some applications within a single laboratory, discipline becomes a necessity when methods have to be transferred reliably between laboratories in an organisation. When the scope of operation encompasses different organisations, national boundaries, *etc.*, a uniformity of approach is essential if comparable results are to be obtained.

This discipline does not come easily, as it requires a control framework. The framework may be considered irksome and unnecessary by some analytical chemists, particularly those from a research environment. It is hoped to persuade those who doubt its necessity that the successful deployment of a method and its wide application rely heavily on such an approach and that flair and technical excellence alone are insufficient.

The foundations for the confidence in an analytical result require that

- the sample is representative and homogeneous;
- the method selected is based upon sound scientific principles and has been shown to be robust and reliable for the sample type under test;
- the instrumentation used has been qualified and calibrated;
- a person who is both competent and adequately trained has carried out the analysis;
- the integrity of the calculation used to arrive at the result is correct and statistically sound.

This guide is concerned with establishing a control framework for the development and validation of laboratory-based analytical methods. Many of these methods will be employed in generating data that could have profound legal or commercial impacts. The validity of analytical results should be established beyond reasonable doubt.

Validation of an analytical method is not a single event. It is a journey with a defined itinerary and stopping places as well as a final destination.

The goal is a method that satisfies the original intent. A disciplined route is

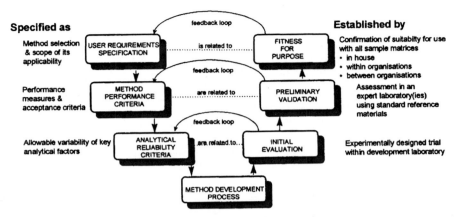

Figure 1 *ISO 'V' model adapted for analytical method validation*

required which maps out the validation journey, more frequently called the validation process.

The ISO 'V' model for system development life cycle in computer software validation is a structured description of such a process. In this instance, the basic 'V' model has been adapted for analytical method validation and is shown in Figure 1.

Like all models, there are underlying assumptions. The main ones for analytical method validation include the areas of equipment qualification and the integrity of the calibration model chosen. If the raw analytical data are produced by equipment that has not been calibrated or not shown to perform reliably under the conditions of use, measurement integrity may be severely compromised. Equally, if the calibration model and its associated calculation methods chosen do not adequately describe the data generated then it is inappropriate to use it. These two areas are considered in some detail in Chapter 8.

Each layer of the ISO 'V' model is dependent upon the layer below and represents stages in the process. Broadly speaking, the boxes in the left-hand portion of the 'V' model represent the aims and objectives of the validation. The boxes in the right-hand portion of the 'V' model contain the processes and procedures that must be carried out successfully and be properly documented to demonstrate that these specified aims and objectives have been met. At the fulcrum of the model is the development process itself.

At each level of the model there is a horizontal correspondence between the two boxes. Verification of the matching of these pairs provides a method of closing the loop at each level.

For example, at the highest level, conformance to the user requirements specification may be verified through data generated in house, through limited laboratory trials or through use of the full IUPAC harmonised protocol. What is critical here is the confirmation of the original user requirements under appropriate performance conditions (Figure 2).

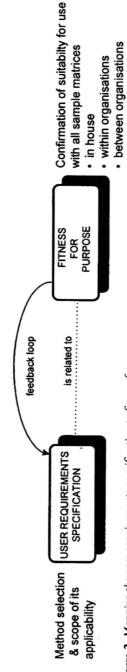

Figure 2 *Mapping the user requirements specification to fitness for purpose*

One useful approach to visualising these relationships is to list bullet points for each of the pairs in the manner shown below. In this way key areas are identified although there are not corresponding relationships between individual bullet points. Individual elements of the model are covered more fully in Chapter 7 where method validation is considered as a whole.

Specified as
- Method applicability
- Analytes to be quantified
- Ranges or limits specified
- Methodology to be used
- Sampling considerations
- Matrices to be covered

etc.

Established by
- Selectivity/specificity
- Linearity
- Accuracy
- Repeatability
- Within-laboratory repeatability
- Reproducibility
- Recovery
- Robustness

etc.

Chapter 8 outlines basic aspects of data evaluation and manipulation. The important topic of linear calibration models is covered in some detail.

Recommended procedures for comparing methods and for taking a single method through to a full IUPAC collaborative trial with the harmonised protocol are covered in Chapter 9. Chapter 10 is a bibliography of recommended books and papers that should be consulted for more details in specific areas.

2 Nomenclature: Terms and Parameters

2.1 Introduction

To avoid confusion, the terms and parameters used in the validation of methods, for example, as used in Figure 3, must be clearly and unambiguously defined. This glossary contains the recommended definitions and corresponding descriptions and is based on the various standards and publications summarised in the Bibliography.[1] This is not exhaustive and it is recommended that the IUPAC 'Orange Book'[2] be consulted if required.

2.2 Terms

2.2.1 Analyte

Component or group of components of which the presence/absence or mass fraction/concentration is to be determined in the test sample.

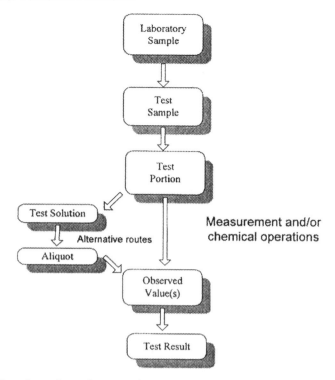

Figure 3 *Flow-chart of sample nomenclature*

2.2.2 Analysis

The method used in the detection, identification and/or determination of the analyte in a sample.

2.2.3 Laboratory sample

The sample or sub-sample(s) of the bulk of the material under consideration sent to or received by the laboratory.

2.2.4 Test sample

A representative quantity of material, obtained from the laboratory sample which is representative for the composition of the laboratory sample.

2.2.5 Test portion

The representative quantity of material of proper size for the measurement of the concentration or other property of interest, removed from the test sample, weighed and used in a single determination.

2.2.6 Observed value

The result of a single performance of the analysis procedure/method, starting with one test portion and ending with one observed value or test result. Note that the observed value may be the average of several measured values on the test portion (2.2.5) *via* the test solution (2.2.8) or aliquots (2.2.9).

2.2.7 Test result

The result of a complete test (frequently a combination of observed values).

2.2.8 Test solution

The solution resulting from dissolving the test portion and treating it according to the analytical procedure. The test solution may be used directly to determine the presence/absence or the mass fraction or mass concentration of the analyte without attributable sampling error. Alternatively, an aliquot (2.2.9) may be used.

2.2.9 Aliquot

A known volume fraction of the test solution (2.2.8) used directly to determine the presence/absence or the mass fraction/concentration of the analyte without attributable sampling error.

2.2.10 Detection

The determination of the presence of the analyte as a chemical entity.

2.2.11 Determination (quantification)

The determination of the absolute quantity of the analyte (mass, volume, mole) or the relative amount of the analyte (mass fraction, mass concentration) in the test sample.

2.2.12 Content mass fraction

The fraction of the analyte in the test sample. The mass fraction is a dimensionless number. However, the mass fraction is usually reported as a quotient of two mass-units or mass-volume.

Value	Mass fraction (SI units)	Non SI units
10^{-2}	% (m/m or m/v)	
10^{-3}	mg g^{-1}, mg mL^{-1}, g kg^{-1}, g L^{-1}	
10^{-4}		
10^{-5}		
10^{-6}	µg g^{-1}, µg mL^{-1}, mg kg^{-1}, mg L^{-1}	ppm, parts per million
10^{-7}		
10^{-8}		
10^{-9}	ng g^{-1}, ng mL^{-1}, µg kg^{-1}, µg L^{-1}	ppb, parts per billion
10^{-10}		
10^{-11}		
10^{-12}	pg g^{-1}, pg mL^{-1}, ng kg^{-1}, ng L^{-1}	ppt, parts per trillion

2.2.13 Mass concentration

The concentration expressed as the mass of the analyte in the test solution divided by the volume of the test solution. The term mass fraction should be used if the amount of the analyte is related to the mass of the sample.

2.1.14 Matrix

All components of the test sample excluding the analyte.

2.3 Parameters

2.3.1 Standard deviation(s)

A measure of the spread in the observed values as a result of random errors (2.3.12). These observed values all have the same expected value. The equation to be used is

$$s = \sqrt{\frac{1}{n-1}\sum_{i=1}^{n}(x_i - \bar{X})^2} = \sqrt{\frac{1}{n-1}\left[\sum_{i=1}^{n}x_i^2 - \frac{1}{n}\left(\sum_{i=1}^{n}x_i\right)^2\right]} \qquad (1)$$

in which x_i = individual measured value, $\bar{X}$ = mean measured value, n = number of measurements.

Equation (1) applies to the calculation of s_r (2.3.7), s_{R_w} (2.3.8) and s_R (2.3.9) under the measurement conditions specified therein.

2.3.2 Relative standard deviation(s) (RSD)

The standard deviation(s) expressed as a percentage of the mean value. The relative standard deviation is defined as:

$$RSD = \frac{s}{X} 100\% \qquad (2)$$

2.3.3 Detection limit

The calculated amount of the analyte in the sample, which according to the calibration line, corresponds to a signal equal to three times the standard deviation of 20 representative blank samples. A blank sample is a sample which does not contain the analyte.

If the recovery (2.3.21) of the analyte is less than 100%, ideally the detection limit should be corrected for the average recovery. However, this is a contentious issue and needs to be considered carefully for each method.

2.3.4 Limit of quantification

The minimum content of the analyte in the test portion that can be quantitatively determined with a reasonable statistical confidence when applying the analytical procedure.

- Report the limit of quantification either in absolute quantities of the analyte (mass, volume or mole) relative amount of the analyte [mass fraction (2.2.12) or mass concentration; (2.2.13)].
- The amount of test portion (for example in grams) must be reported as used in the determination.

The limit of quantification is numerically equivalent to six times the standard deviation of the measured unit when applying the analytical procedure to 20 representative blank samples. For recoveries less than 100% the limit of quantification must be corrected for the average recovery of the analyte.

2.3.5 Sensitivity

The change of the measured signal as a result of one unit change in the content of the analyte.

The change is calculated from the slope of the calibration line of the analyte.

2.3.6 Rounding off

The process of achieving agreement between an observed value and the repeatability (2.3.7) of the analytical procedure. The maximum rounding off interval is equal to the largest decimal unit determined to be smaller than half the value of the standard deviation of the repeatability (2.3.7). See Section 8.3.1 for more details.

2.3.7 *Repeatability (r)*

The expected maximum difference between two results obtained by repeated application of the analytical procedure to an identical test sample under identical conditions.

The measure for repeatability (r) is the standard deviation (s_r). For series of measurements of a sufficient size (usually not less than 6), the repeatability is defined as

$$r = 2.8 \times s_r \text{ (confidence level 95\%)} \tag{3}$$

Repeatability should be obtained by the same operator with the same equipment in the same laboratory at the same time or within a short interval using the same method.

2.3.8 *Within-laboratory reproducibility (R_w)*

The expected maximum difference between two results obtained by repeated application of the analytical procedure to an identical test sample under different conditions but in the same laboratory. The measure for the within-laboratory reproducibility (R_w) is the standard deviation (s_{R_w}).

For series of measurements of sufficient size (usually not less than 6), the within-laboratory reproducibility is defined as

$$R_w = 2.8 \times s_{R_w} \text{ (confidence level 95\%)} \tag{4}$$

Within-laboratory reproducibility should be obtained by one or several operators with the same equipment in the same laboratory at different days using the same method.

2.3.9 *Reproducibility (R)*

The expected maximal difference between two results obtained by repeated application of the analytical procedure to an identical test sample in different laboratories. The measure for the reproducibility (R) is the standard deviation (s_R).

For series of measurements of sufficient size (usually not less than 6) the reproducibility is defined as

$$R = 2.8 \times s_R \text{ (confidence level 95\%)} \tag{5}$$

Between-laboratory reproducibility should be obtained by different operators with different instrumentation in different laboratories on different days using the same method.

For a given method, the most important factors in the determination of repeatability and reproducibility are Laboratory, Time, Analyst and Instrumentation.

Experimental condition to determine	Factors to vary or control
Repeatability	Same L, T, A, I
Within-laboratory reproducibility	Same L; different T; I and A may be different
Between-laboratory reproducibility	Different L, T, A, I

If it is not possible to involve additional laboratories for the determination of the between-laboratory reproducibility, then the within-laboratory reproducibility may be used to get an estimate of the between-laboratory reproducibility. The reproducibility of the method may be dependent upon the mass fraction of the analyte in the test sample. It is therefore recommended, when studying the reproducibility, to investigate whether a relation exists between concentration and reproducibility. The measurement series should be greater than 8.

2.3.10 Trueness

The closeness of agreement between the average value obtained from a large series of test results and an accepted reference value. The measure of trueness is usually expressed in terms of bias.

2.3.11 Systematic error or bias

The difference between the average observed value, obtained from a large series of observed values ($n \geq 8$), and the true value (2.3.13) (Figure 4).

2.3.12 Random error

The difference between a single observed value and the average value of a large number of observed values (at least 8), obtained by applying the same analytical procedure to the same homogeneous test sample.

2.3.13 True value

The value that describes the content and is completely defined by the circumstances under which the content has been determined.

2.3.14 Precision

A measure of the agreement between observed values obtained by repeated application of the same analytical procedure under documented conditions (2.3.7–2.3.9).

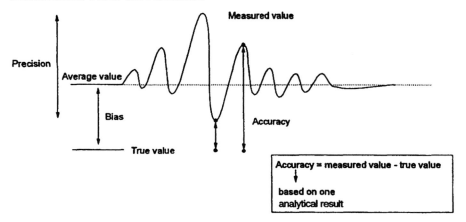

Figure 4 *Accuracy, precision and bias*

2.3.15 Accuracy

A measure of the agreement between a single analytical result and the true value.

2.3.16 Ruggedness

Ruggedness of an analytical method is the insensibility of the method for variations in the circumstance and the method variables during execution. It is important to note that statistically significant deviations are not always relevant. The purpose for which the measurement is made is a more important criterion for deciding on the ruggedness and the statistical method employed is merely a tool.

2.3.17 Noise

A phenomenon defined as fast changes in the intensity and frequency of a measured signal irrespective of the presence or absence of the analyte. The speed of change is significantly different from the normally expected detector response. A measure of noise is the measured difference between the highest and lowest value of the measured signal with no analyte present, observed in a relatively short time-span, as compared to the time-span necessary for measurement of the analyte.

2.3.18 Selectivity

A measure of the discriminating power of a given analytical procedure in differentiating between the analyte and other components in the test sample.

2.3.19 Significant figures

Values which contain the information consistent with either the repeatability or reproducibility of the analytical procedure. Significant values are obtained by using the described method for rounding off (Section 8.3.1).

2.3.20 Specificity (see also Selectivity)

The property of the analytical procedure to measure only that which is intended to be measured. The method should not respond to any other property of the analyte or other materials present.

2.3.21 Recovery

The fraction of the analyte determined in a blank test sample or test portion, after spiking with a known quantity of the analyte under predefined conditions.

- Recovery is expressed as a percentage.
- The part of the analytical procedure in which recovery is involved should be reported.
- The critical stages/phases relating to instability, inhomogeneity, chemical conversions, difficult extractions, *etc.* should be reported.
- Recovery must not be based on an internal standard unless work is undertaken to demonstrate identical behaviour under the conditions of test.

2.3.22 Scope

The collection of matrices and analyte(s) to which the analytical procedure is applicable.

2.3.23 Range

The content mass fraction interval to which the analytical procedure is applicable.

2.3.24 Drift

The phenomenon observed as a continuous (increasing or decreasing) change (slowly in time) of the measured signal in the absence of the analyte.

3 Samples and Sampling

3.1 Introduction

The importance of sampling in method validation and, in particular, inter-comparison of methods cannot be overemphasised. If the test portion is not representative of the original material, it will not be possible to relate the analytical result measured to that in the original material, no matter how good the analytical method is nor how carefully the analysis is performed. It is essential that the laboratory sample is taken from a homogeneous bulk sample as a collaborator who reports an outlying value may claim receipt of a defective laboratory sample. It is important to understand that sampling is always an error generating process and that although the reported result *may* be dependent upon the analytical method, it will *always* be dependent upon the sampling process.

The essential question in the inter-comparison of analytical methods is, 'If the same sample (or a set of identical aliquots of a sample) is analysed by the same method in different laboratories, are the results obtained the same within the limits of experimental error?'. It is apparent, therefore, that the selection of an appropriate sample or samples is critical to this question and that the sampling stage should be carried out by a skilled sampler with an understanding of the overall context of the analysis and trial.

Any evaluation procedure must cover the range of sample types for which the method under investigation is suitable, and details of its applicability in terms of sample matrix and concentration range must be made clear. Similarly, any restrictions in the applicability of the technique should be documented in the method.

For more details, the works listed in the Bibliography should be consulted. In particular, Crosby and Patel's *General Principles of Good Sampling Practice*[3] and Prichard[4] provide readily digestible guidance to current best practices in this area.

3.2 What is a sample?

The Commission on Analytical Nomenclature of the Analytical Chemistry Division of the International Union of Pure and Applied Chemistry has pointed out that confusion and ambiguity can arise around the use of the term 'sample' and recommends that its use is confined to its statistical concept. When being used to describe the material under analysis, the term should be qualified by the use of 'laboratory sample' or 'test sample', for example.

One of the best treatments of sampling terminology is given in recommendations published by IUPAC[5] which describes the terms used in the sampling of bulk or packaged goods. In this example, the sampling procedure reduces the original *consignment* through *lots* or *batches, increments, primary* or *gross samples, composite* or *aggregate samples, subsamples* or *secondary samples* to a *laboratory sample*. The *laboratory sample*, if heterogeneous, may be further prepared to produce the *test sample*. Arrival at either a laboratory sample or test sample is deemed to be the end of the sampling procedure.

Once received into the laboratory the laboratory samples or test samples will be recorded and then be subjected to analytical operations, beginning with the measuring out of a test portion and proceeding through various operations to the final measurement and reporting of results/findings.

The IUPAC nomenclature for the sampling process is illustrated in Figure 5. This links with the sampling nomenclature diagram on Page 8 (Figure 3).

The problems associated with sampling in many areas of chemical testing have been addressed and methods have been validated and published (see ref. 3 for more details). Where specific methods are not available, the analytical chemist should rely upon experience or adapt methods from similar applications. When in doubt, the material of interest and any samples taken from it

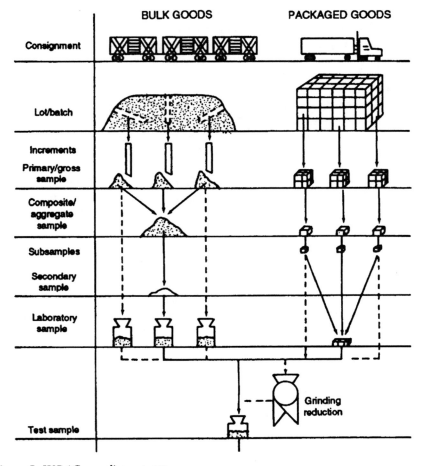

Figure 5 *IUPAC sampling process*

should always be treated as heterogeneous. It is important when documenting a sampling procedure to ensure that all of the terms are clearly defined, so that the procedure will be clear to other users. The use of sampling plans may be appropriate and guidance is available for procedures based upon attributes or variables.[6]

3.3 Homogeneity and concentration ranges

Extreme care must be taken to ensure that the bulk sample from which the laboratory or test samples are taken is stable and homogeneous—this is particularly important if 'spiked' samples are provided.

The homogeneity should be established by testing a representative number of laboratory samples taken at random using either the proposed method of analysis or other appropriate tests such as UV absorption, refractive index, *etc.* The penalty for inhomogeneity is an increased variance in analytical results that is not due to intrinsic method variability.

For quantitative analysis the working range for a method is determined by examining samples with different analyte concentrations and determining the concentration range for which acceptable accuracy and precision can be achieved. The working range is generally more extensive than the linear range, which is determined by the analysis of a number of samples of varying analyte concentrations and calculating the regression from the results (see Section 8.2 for more details). For a comprehensive study, which has been designed to evaluate the method fully, samples possessing low, medium and high concentration levels of the analyte to be determined must be prepared. The only exception to this would be when the level of the analyte always falls within a narrow range of concentrations.

4 Method Selection

4.1 'Fitness for purpose'

Far too often, method selection is carried out by deciding to apply the technique that is most popular or familiar. If a laboratory has expertise in a particular technique then it is tempting to let that expertise be the overriding factor in method selection. Rarely is there a structured and considered approach to method selection. Whilst it is often possible to make inappropriate methods work within a single laboratory, the impact on the reliable transfer between laboratories can be very large. In the past, the transferability of methods has not been given the prominence it deserves. However, within the current climate of harmonisation and interchangeability, the technical requirements of method transfer and method performance have been addressed in some detail and are covered in Chapter 9. There are two areas which have received less attention and agreement, namely the inter-comparison of different methods for the same analytes in-house or within a few laboratories and the methods for describing and writing analytical methods. The former topic is the subject of Section 9.3.

The latter is discussed in Section 7.2. No method is 'fit for purpose' unless there are clear and unambiguous written instructions for carrying out the prescribed testing in accordance with the conditions laid down in the original method development cycle.

The literature contains examples of collaborative trials that only prove that the method was not fit for its intended purpose! The full IUPAC harmonised protocol is by its very nature an extensive and expensive exercise. From an economic perspective such trials should only be undertaken when there is good and well-documented evidence that it is likely that the method under evaluation is sufficiently robust. Investment of time and intellectual effort in method selection and the other aspects of the user requirements specification will pay great dividends. Prevention is better and nearly always cheaper than cure.

4.2 Sources and strategies

Once the User Requirements Specification has been drawn up and the method performance criteria set, the method development process can begin. Quite often there are existing methods available within the literature or within trade and industry. On many occasions it is tempting to ignore the difficulties of a comprehensive literature search to save time. However, as a minimum, key word searches through the primary literature and abstracting journals such as *Analytical Abstracts* and *Chemical Abstracts* should be undertaken. For standard or statutory methods, it is essential to scan international standards from Europe and the USA as well as local sources and those deriving from statutory publications. Once existing methods have been identified, it is good practice to compare them *objectively*. One way to do this is to list the performance criteria and relevant sections of the User Requirements Specification and tabulate the corresponding data.

An existing method may have a sufficiently good fit that adaptation is likely to lead to a suitable method. This relies upon professional knowledge and experience.

For methods that are likely to be widely used, other aspects of suitability need to be considered.

Some areas for consideration are listed below.

- Can the method be written down sufficiently clearly and concisely to allow ease of transfer?
- Can all the critical method parameters be identified and controlled? This is particularly important where automated systems are involved.
- Is the equipment readily available to all the likely participants? This assumes a special importance for internationally distributed methods and may involve questions of maintenance and support.
- Are all the reagents and solvents readily available in the appropriate quality?
- Do the staff have the requisite skills and training to carry out the procedure?

- Are Health and Safety or environmental considerations likely to cause problems?
- Are standards and reference materials readily available to ensure that equipment and systems are properly calibrated and qualified?

4.3 Sampling considerations

In the enthusiasm for a particular technique or method, it is sometimes the case that the appropriateness of sample size is overlooked. Even though the sampling process outlined in Chapter 3 has been followed, it is essential that the size of the test sample and its relationship to the sizes of the test portion and the test solution are considered. These factors need to be considered during method selection.

For example, is a 2 g test sample from a 1000 kg consignment or bulk batch adequate for the purpose? It may be under appropriate circumstances. If not, how much material needs to be taken? Recent draft guidance to the pharmaceutical industry from the FDA[7] recommends that for blend uniformity sample sizes no more than three times the weight of an individual dose should be taken.

Equally, consideration needs to be given to sample presentation. Is it more appropriate to test non-destructively to gain physical and chemical information or by solution/extraction processes to separate the analyte(s) of interest?

The most important aspect here is that these questions have been asked and documented answers given as part of the User Requirements Specification.

It is essential to remember that whilst any test result may be method-dependent it is always sample-dependent.

4.4 Matrix effects

As far as is practically possible, the selection and preparation of samples must take into account all possible variations in the matrix of the material to be analysed. The applicability of the method should be studied using various samples ranging from pure standards to mixtures with complex matrices as these may contain substances that interfere to a greater or lesser extent with the quantitative determination of an analyte or the accurate measurement of a parameter. Matrix effects can both reduce and enhance analytical signals and may also act as a barrier to recovery of the analyte from a sample.

Where matrix interferences exist, the method should ideally be validated using a matched matrix certified reference material. If such a material is not available it may be acceptable to use a sample spiked with a known amount of the standard material.

The measurement of the recoveries of analyte added to matrices of interest is used to measure the bias of a method (systematic error) although care must be taken when evaluating the results of recovery experiments as it is possible to obtain 100% recovery of the added standard without fully extracting the analyte which may be bound in the sample matrix.

The whole question of recovery adjustment is a vexed one. In theory, one should always carry out this correction. However, the best analytical practice is to consider the question for each application and sample matrix combination and make and document the decision. For more detailed information, the recently published 'Harmonised Guidelines for the Use of Recovery Information in Analytical Measurements'[8] should be consulted.

5 Equipment Calibration and Qualification

Analytical practitioners place great faith in the readings and outputs from their instruments. When unexpected or out of specification results occur, the initial suspicion often falls on the sample, the preparation technique or the analytical standard employed. Rarely is the equipment questioned. Indeed, the whole underpinning of method validation assumes that the analytical equipment used to acquire the experimental data is operating correctly and reliably.

Some industries that are highly regulated, such as the pharmaceutical sector, have placed great emphasis on method validation in, for example, HPLC.[9,10] However, until recently, there has been little direct requirement for assuring that the analytical instruments are working properly.

The major regulatory guidelines for Good Manufacturing Practice (GMP) and Good Laboratory Practice (GLP) are similarly vague. 'Fitness for purpose' is the phrase that is commonly used, but what does this mean in practice? Primarily, the Pharmacopoeias[11,12] and the Australian Regulatory Authority[13] have been sufficiently worried by instrumental factors to give written requirements for instrument performance. Whilst these guidelines are not consistent at least they are attempting to ensure consistent calibration practices between laboratories.

In contrast, the ISO Guide 25 approach (updated in 1999 to ISO Guide 17025), as expanded in ref. 14 heavily focuses on good analytical practices and adequate calibration of instruments with nationally or internationally traceable standards wherever possible.

5.1 Qualification approaches

Equipment qualification is an essential part of quality assuring the analytical data on which our knowledge of the sample rests. The importance of this 'data to information iceberg' is illustrated in Figure 6. There are several approaches commonly employed.

5.1.1 The 'bottom up' approach

The 'bottom up' approach ensures the quality of the end result by building up from the foundations rather like a *Lego* model. In testing terms this is illustrated in Figure 7. These *Lego* bricks are equivalent to the individual modules in any measurement system. Each brick is qualified or confirmed as suitable for use before the next layer is built. In this way, integrity is assured all the way to the

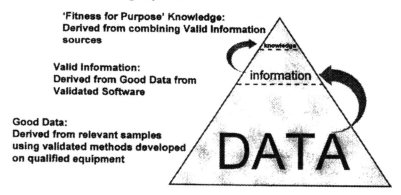

'Fitness for Purpose' Knowledge:
Derived from combining Valid Information
sources

Valid Information:
Derived from Good Data from
Validated Software

Good Data:
Derived from relevant samples
using validated methods developed
on qualified equipment

Figure 6 *The 'data to information' iceberg*

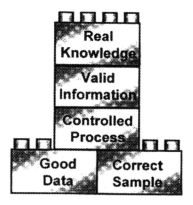

Figure 7 *The 'bottom up' approach*

top-most layer. If firm foundations are not built, the information generated will not stand scrutiny. By following this approach quality is built in from the lowest level.

The role of the instrument in providing the integrity of data is fundamental to the end result. If the analytical practitioner cannot have faith in the reliability of the basic analytical signal within predetermined limits then the information generated will be worse than useless. Reliability of the data quality should be linked to performance standards for both modules and systems as well as having a regular maintenance programme.

5.1.2 The 'top down' approach

An alternative and increasingly applied approach, particularly from the regulatory bodies, is from the other direction, *i.e.* 'top down'. This approach is known as the 4Qs model, **DQ, IQ, OQ** and **PQ** which are:

Design Qualification
Installation Qualification
Operational Qualification and
Performance Qualification

A detailed discussion of this approach may be found in refs. 15–19 and references therein. By way of example, however, the approach will be illustrated with respect to a typical analytical instrument.

Design Qualification, **DQ**, is about specifying what the instrument or instrument system has to do. This would include documenting technical requirements, environmental conditions, sample and sample presentation requirements, data acquisition and presentation needs, operability factors and any Health & Safety issues. In addition a cost–benefit analysis would normally be performed.

The Instrumental Criteria Sub-committee of the Analytical Methods Committee has been active for many years in producing Guidelines for the Evaluation of Analytical Instrumentation. Since 1984, they have produced reports on atomic absorption, ICP, X-ray spectrometers, GLC, HPLC, ICP-MS, molecular fluorescence, UV–Vis–NIR, IR and CE. These are excellent source documents to facilitate the equipment qualification process. A current listing of these publications is given in Section 10.2.

Having chosen the analytical instrument or system, Installation Qualification, **IQ**, should be carried out to ensure that the equipment works the way the vendor or manufacturer specifies it should. **IQ** should be performed in accordance with a written test protocol with acceptance criteria with certification from the installation engineer, who is suitably qualified. Full written records of all testing carried out should be maintained as well as ensuring that adequate documentation and manuals have been supplied. The latter should include any Health & Safety information from vendor or manufacturer.

Once satisfied that the instrument is operating in accordance with its own specification, the end user should ensure that it is 'fit for purpose' for the applications intended. This step is called Operational Qualification, **OQ**. This process would include writing the Standard Operating Procedure (SOP) and training staff in its use. Further testing may be required to ensure that the instrument performance is in accordance with National and Corporate standards if not carried out in **IQ**. Frequently, instruments are used with accessories or sub-systems, *e.g.* sipper systems or other sample presentation devices. Challenge the analytical system with known standards and record what you did. It is necessary to ensure that they work in the way intended and that documented evidence is available to support their use.

Calibration procedures and test methods and frequencies need to be defined usually as part of an SOP. If you intend to transfer data from the instrument to a software package, ensure that data integrity is preserved during transfer. Don't assume that the transfer protocols on 'standard' interfaces always work as intended. It is good practice to ensure that the data have not been truncated or distorted during transfer.

At this point in the process, the equipment/system is able to be put into routine use. The final **Q** in the model, Performance Qualification, **PQ**, is about on-going compliance. Elements of **PQ** include a regular service programme, performance monitoring with warning and action limits (as defined in **OQ**). All of these elements need to be documented and individual log books for systems are useful for this purpose. **PQ** data should be subject to regular peer review. All instrument systems should be subject to a simple change procedure which may well be connected to the equipment log system.

5.1.3 Holistic approach

Furman *et al.*,[17] discussing the validation of computerised liquid chromatographic systems, present the concept of modular and holistic qualification. Modular qualification involves the individual components of a system such as pump, autosampler, column heater and detector of an HPLC. The authors make the point that:

'calibration of each module may be useful for trouble shooting purposes, such tests alone cannot guarantee the accuracy and precision of analytical results'.

Therefore the authors introduced the concept of holistic validation where the whole chromatographic system was also qualified to evaluate the performance of the system. The concept of holistic qualification is important as some laboratories operate with a policy of modular equipment purchase. Here they select components with the best or optimum performance from any manufacturer. Furthermore, some of these laboratories may swap components when they malfunction. Thus, over time the composition of a system may change. Therefore, to assure themselves and any regulatory bodies that the system continues to function correctly, holistic qualification is vital.

Most laboratory systems require maintenance and inclusion preventative maintenance programmes. Therefore any holistic testing should form part of Performance Qualification to ensure on-going compliance.

5.2 A convergence of ideas

Much in the way of harmonisation of procedures and practices in analytical chemistry has been going on outside these activities. Many of these initiatives are now coming to fruition. CITAC (Co-operation on International Traceability in Analytical Chemistry) have produced an *International Guide to Quality in Analytical Chemistry*[18] which attempts to harmonise the following regulatory codes of practice for the analytical laboratory: ISO Guide 25 (revised in December 1999 to ISO Guide 17025), ISO 9001 and 9002 and GLP. A VAM Instrumentation Working Group has published *Guidance on Equipment Qualification of Analytical Instruments: High Performance Liquid Chromatography (HPLC)*.[19]

If there is one compelling reason for equipment qualification, it lies within the need to transfer methods between laboratories. Why are so many of our collaborative trials a failure? One answer lies in the fact that the key analytical variables are not always identified and controlled through specification and/or procedural practice. These may lie within the method but more often are due to the operating parameters of the equipment or system. If, for example, temperature is a key factor, how can it be specified if there is no assurance that instrument A's temperature readout is operating within known accuracy and precision limits? Furthermore, if a laboratory is transferring a method involving an HPLC gradient separation and there is no equipment specification at the level of the pump, there may be problems in the technology transfer. Consideration needs to be given to the effects of choosing high pressure versus low pressure solvent mixing and differences in dead volume between the pump and column which can affect the gradient formation. These factors are likely to affect the quality of the separation achieved. Without specification there can be no reliable control. Another reason may be that the overall analytical process capability is affected by one or more instrumental factors. Methods developed on unqualified equipment or systems may well lack the robustness and reliability needed.

Calibration is often confused with qualification. As pointed out by Parriott[20] with reference to HPLC methods:

'The term calibration implies that adjustments can be made to bring a system into a state of proper function. Such adjustments generally cannot be performed by chromatographers and are best left to trained service engineers who work for, or support, the instrument manufacturers.'

Calibration is, therefore, inextricably linked to equipment qualification and preventative maintenance. Whenever calibration involves adjustments of the type described above, it is important to document the activity and where appropriate re-qualify the instrument concerned.

6 The Method Development Process

The overall process from concept to validated method is illustrated on Page 4 (Figure 1). Once an appropriate analytical principle has been selected and the method performance criteria defined, the actual method development process can begin. Usually, this phase is carried out using pure materials and limited samples that are known, or assumed, to be homogeneous.

The purpose of this process is to confirm the viability of the method chosen and show that the procedure is sufficiently analytically robust to allow a preliminary validation to be carried out. The AOAC collaborative study guidelines[21] explicitly state

'Do not conduct collaborative study with an unoptimized method. An unsuccessful study wastes a tremendous amount of collaborators' time and

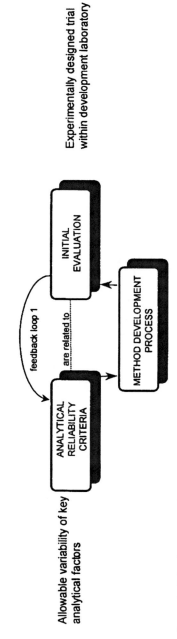

Figure 8 *Method development cycle*

creates ill will. This applies especially to methods formulated by committees and have not been tried in practice.'

The key factors that need to be established at this stage include:

- applicability of the analytical principle(s) over the concentration range required;
- optimisation of experimental conditions;
- selection of the calibration function;
- selection of reference materials and standards;
- evaluation of matrix effects and interferences;
- recovery experiments;
- robustness of the procedure to changes in key parameters;
- generation of initial accuracy and precision data.

As is indicated in Figure 8, this process is likely to be an iterative one. However, it is essential that good written records are kept during this phase so that, in the event of problems at subsequent levels, investigations may be more readily carried out. Alas, far too often the excuse of 'analytical creativity' is cited for lack of such records. The most important outcome from this initial evaluation should be an assessment of robustness (or ruggedness) of the developed procedure. The AOAC Guide,[22] *Use of statistics to develop and evaluate analytical methods* is an excellent source for a discussion of statistical procedures for both inter- and intra-laboratory studies.

Recently, the topic of method development for both routine and non-routine analyses has been the subject of two EURACHEM documents; *The Fitness for Purpose of Analytical Methods*[23] and *Quality Assurance for Research and Development and Non-routine Analysis*[24] as part of the VAM (Valid Analytical Measurements) programme. These guides provide information and a bibliography for ISO publications.

6.1 Mapping the analytical process and determining the key factors

The identification of the key factors involved is crucial in planning the development process (Figure 9). Consideration needs to be given to each of the analytical issues and the outcome documented. A well-written laboratory notebook is essential in recording such information in a timely manner. Efforts expended here will greatly facilitate the writing of the finalised analytical method. For example, the basic assumptions regarding recovery and selectivity issues may have a profound effect on the detailed description of the sample workup procedure. The other areas which are often under-specified are assuring the integrity of data transfer and transformation.

If spreadsheets are to be used it is prudent to ensure that any macros and procedures are correct and that the in-built statistical functionality is appropriate! It is very easy to select the s_n function instead of s_{n-1}. Remember that s_n

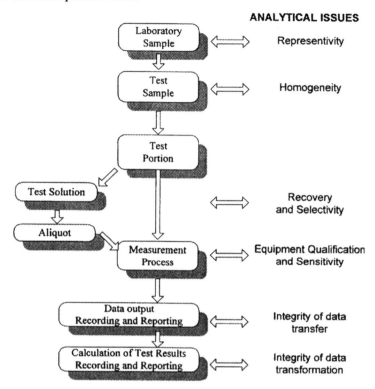

ANALYTICAL ISSUES

Laboratory Sample ⟵⟶ Representivity

Test Sample ⟵⟶ Homogeneity

Test Portion

Test Solution

Aliquot

Recovery and Selectivity

Measurement Process ⟵⟶ Equipment Qualification and Sensitivity

Data output Recording and Reporting ⟵⟶ Integrity of data transfer

Calculation of Test Results Recording and Reporting ⟵⟶ Integrity of data transformation

Figure 9 *Mapping the analytical process*

refers to the entire population and s_{n-1} to a sample from a population. This applies also to hand-held calculators.

A procedure for determining which factors are important is to use Sandel's Venn diagram approach.[25] An adapted form is shown in Figure 10. Wernimont[22] has developed this idea for intra-laboratory studies. Note, however, that each of the three factors may be affected by external events.

The purpose of the development process is to determine the contributory variances to each of these three areas in order to minimise or control them. The instrument performance needs to be assured and this has been discussed in Chapter 5. Even if we assume initially that the operator contribution is small, we need to confirm that during this phase. Trust but verify!

6.2 Simple experimental design

A simple example, focusing on the analytical procedure, will illustrate the type of experimental design used to investigate three key factors in an HPLC method. Detailed discussion of experimental designs for robustness testing can be found in Morgan[26] and Hendriks *et al.*[27] Riley and Rosanske[28] provide an

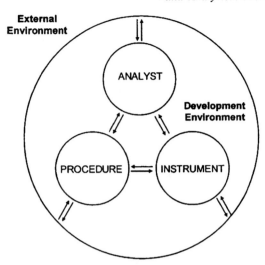

Figure 10 *Sandel's Venn diagram for method development*

overview from a USA pharmaceutical perspective. For those wishing to go deeply into the subject, Deming and Morgan,[29] Montgomery[30] and Box and Hunter[31] are more detailed sources.

Consider an HPLC method for the separation of 11 priority pollutant phenols using an isocratic system. The aqueous mobile phase contains acetic acid, methanol and citric acid. From preliminary studies, it was established that the mobile phase composition was critical to ensure maximum resolution and to minimise tailing. The overall response factor, CRF, was measured by summing the individual resolutions between pairs of peaks. Hence, the CRF will increase as analytical performance improves.

The data for this example are taken from ref. 26 in the bibliography. Many experimental designs are available but a simple full factorial is taken by way of example. A full factorial design is where all combinations of the factors are experimentally explored. This is usually limited from practical consideration to low values. To simplify the matter further no replication was used.

The design chosen is a full factorial 2^3 with two levels of each of the three factors, acetic acid concentration, methanol concentration and citric acid concentration. The low ($-$) and high ($+$) levels of each are shown in Table 1.

Table 1 *Mobile phase factors for the two level full factorial 2^3 design*

Factor		Low (−)	High (+)
Acetic acid concentration (mol dm^{-3})	A	0.004	0.010
Methanol (%v/v)	M	70	80
Citric acid concentration (g L^{-1})	C	2	6

Table 2 *Experimental design matrix for the two level full factorial 2^3 design*

Run no.	A	M	C
1	0.004	70	2
2	0.010	70	2
3	0.004	80	2
4	0.010	80	2
5	0.004	70	6
6	0.010	70	6
7	0.004	80	6
8	0.010	80	6

These extreme levels were established during the preliminary method development work.

A full factorial 2^3 design allows the study of three main factors and their interactions to be carried out in eight experiments or runs. The first requirement is to set out the experimental design matrix. This is shown in Table 2. 'A' is the molar concentration of the acetic acid, 'M' is the methanol concentration, %v/v, and 'C' is the citric acid concentration, g L^{-1}. All combinations are covered in eight experimental runs. Note that this is not the order in which they are performed. These should be carried out in a random sequence. There will be a value of the CRF for each run.

This design matrix for the main effects may be expressed also in the high/low or $+/-$ notation. The values for the CRF have been added to this and are shown in Table 3.

This design matrix shows only the main effects, *i.e.*, A, M and C. However, the 2^3 design allows their two-factor interactions, AM, AC and AC, to be calculated as well as one of the three-factor interactions AMC. It is unlikely that a three-factor interaction will be significant although in some instances two-factor interactions are important.

Table 3 *Experimental design matrix with contrast coefficients and experimental values*

Run no.	A	M	C	CRF
1	−	−	−	10.0
2	+	−	−	9.5
3	−	+	−	11.0
4	+	+	−	10.7
5	−	−	+	9.3
6	+	−	+	8.8
7	−	+	+	11.9
8	+	+	+	11.7

One of the best ways of visualising a two level 2^3 design is to consider a cube with each of the three axes representing one of the factors. Hence, each of the two levels is represented as a value on each axis with the eight vertices having a corresponding experimental result. In this example the experimental result would be the CRF value. This design is shown in Figure 11.

In our example this template can be filled in using the values from Tables 2 and 3 and is shown in Figure 12.

In order to decode the effects and interactions the full design matrix with all the contrast coefficients (columns) is needed. This is shown in Table 4. The 'I' column contains the data of all the CRF values and is used to calculate the overall mean effect.

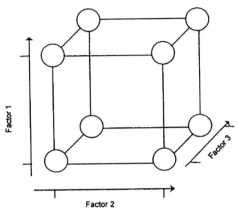

Figure 11 *Visual representation of a two level full factorial 2^3 design*

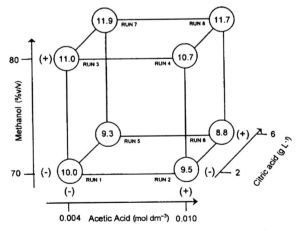

Figure 12 *Visual representation of the two level full factorial 2^3 design for the HPLC example*

Table 4 *Full design matrix for a two level full factorial 2^3 design*

Run no.	I	A	M	AM	C	AC	MC	AMC
1	+	−	−	+	−	+	+	−
2	+	+	−	−	−	−	+	+
3	+	−	+	−	−	+	−	+
4	+	+	+	+	−	−	−	−
5	+	−	−	+	+	−	−	+
6	+	+	−	−	+	+	−	−
7	+	−	+	−	+	−	+	−
8	+	+	+	+	+	+	+	+

The normal way of carrying out such decoding calculations is by use of specialised software or a customised spreadsheet. However, for the purposes of illustration, the calculation process will be described. Basically, all that is required is to superimpose the sign convention of the contrast coefficients onto the experimental responses and perform some simple arithmetic. For this example, the calculations are shown in Table 5.

Each of the columns is summed and divided by the number of data pairs (4) with the exception of the first one which is merely the original CRF values unchanged. Dividing this summation by the number of data values, 8, gives the overall mean effect.

The values for each of the main and interaction effects are listed in Table 5. The larger the absolute magnitude of the value the greater is its importance. The sign of the value indicates the direction of the change of the effect. For method development purposes, we need to know which are the large effects so that they may be controlled. The question is what is large enough to be significant?

Table 5 *Completed table of contrasted values for effect calculations*

Run no.	I	A	M	AM	C	AC	MC	AMC
1	10.0	−10.0	−10.0	10.0	−10.0	10.0	10.0	−10.0
2	9.5	9.5	−9.5	−9.5	−9.5	−9.5	9.5	9.5
3	11.0	−11.0	11.0	−11.0	−11.0	11.0	−11.0	11.0
4	10.7	10.7	10.7	10.7	−10.7	−10.7	−10.7	−10.7
5	9.3	−9.3	−9.3	9.3	9.3	−9.3	−9.3	9.3
6	8.8	8.8	−8.8	−8.8	8.8	8.8	−8.8	−8.8
7	11.9	−11.9	11.9	−11.9	11.9	−11.9	11.9	−11.9
8	11.7	11.7	11.7	11.7	11.7	11.7	11.7	11.7
Sum	82.9	−1.5	7.7	0.5	0.5	0.1	3.3	0.1
Divisor	8	4	4	4	4	4	4	4
Effect	10.363	−0.375	1.925	0.125	0.125	0.025	0.825	0.025
SSQ		0.28125	7.41125	0.03125	0.03125	0.00125	1.36125	0.00125
Total SSQ	9.11875							

Table 6 *Ranking of effects by value and their normal probability*

Rank	Effect		Value	P
1	Acetic acid concentration	A	−0.375	7.14
2	Interaction	AC	0.025	21.43
3	Interaction	AMC	0.025	35.71
4	Interaction	AM	0.125	50.00
5	Citric acid concentration	C	0.125	64.29
6	Interaction	MC	0.825	78.57
7	Methanol concentration	M	1.925	92.86

Ranking (ordering) them in ascending order value is a good place to start (Table 6).

What is clear without the further aid of statistics is that the methanol concentration is the most important factor. Equally, it is clear that the citric acid concentration is not significant nor are three of the four interactions. Are the methanol concentration main effect and/or the interaction between the methanol and citric acid concentrations significant? One way forward is to plot the data from Table 6 on normal probability paper. If all these data are insignificant then they will lie on a straight line. If values are observed that are a long way off the line it is likely that the effects or interactions are significant.

This is easily done because the relationship between the rank of the effect, i, the total number of effects, T, and the expected probability, P, is:

$$P = 100 \frac{(i - 0.5)}{T} \qquad (6)$$

The calculated values listed in Table 6 are plotted in Figure 13. Note that the probability, P, is plotted on a logarithmic scale.

Examining Figure 13, M is clearly way off the line. Also, A does not lie on the line but is unlikely to be significant. The question about MC is more difficult to answer but for the moment it is assumed that it not significant. This issue may be resolved, however, by conducting replicate experiments that provide an independent estimate of the residual error—this will be discussed later.

However, another way of extracting information from these data can be made by conducting an analysis of variance, ANOVA. In Table 7, the sum of squares (SSQ) of each of the effects and also the overall sum of squares have been extracted from Table 5. These data are retabulated in Table 8 in the more usual ANOVA format. Once again, the methanol concentration is a large factor.

The variance ratio (F value) is not readily calculated because replicated data are not available to allow the residual error term to be evaluated. However, it is usual practice to use the interaction data in such instances if the normal probability plot has shown them to be on the linear portion of the graph. By grouping the interaction terms from Table 7 as an estimate of the residual error,

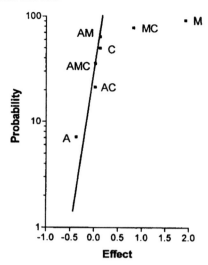

Figure 13 *Normal plot of the effects of the two level full factorial 2^3 design for the HPLC example*

Table 7 *Tabulation of effects data for the two level full factorial 2^3 design for the HPLC example in ANOVA format*

Source of variation	SSQ	DF	MS
Acetic acid concentration (A)	0.28125	1	0.28125
Methanol concentration (M)	7.41125	1	7.41125
Citric acid concentration (C)	0.03125	1	0.03125
Interactions AM	0.03125	1	0.03125
AC	0.00125	1	0.00125
MC	1.36125	1	1.36125
AMC	0.00125	1	0.00125
Total	9.11875	7	

Table 8 *Revised ANOVA table for the two level full factorial 2^3 design for the HPLC example*

Source of variation	SSQ	DF	MS	F value
Acetic acid concentration (A)	0.28125	1	0.28125	0.81
Methanol concentration (M)	7.41125	1	7.41125	21.25
Citric acid concentration (C)	0.03125	1	0.03125	0.09
Residual error	1.395	4	0.34875	
Total	9.11875	7		

we can calculate the variance ratio (F value) by dividing the Mean Square errors of the main effects by the residual Mean Square error. The results are shown in Table 8. From tables of the F distribution, the critical value for 95% confidence for $F(1,4$ df) is 7.71. Therefore the methanol effect is the only significant one.

Suppose that the data used for the CRF values were mean values from a duplicate experiment. Then it would be possible to obtain an estimate of the error by pooling the data. By taking the mean difference squared of the data pairs a run variance is obtained. A pooled estimate is calculated by summing all eight run variances and taking the mean value. This calculation is shown in Table 9.

If it is assumed that the pooled run variance is a reasonable estimate for the residual variance, Table 7 can be reworked and the variance ratios (F values) calculated for each of the effects. The results of this rework are shown in Table 10. This approach confirms that the methanol effect is the largest by a very long way. The F value (1,8 df) is 5.32. Whilst this confirms that A is not significant,

Table 9 *Use of the run variances to generate an estimate of the residual variance*

Run no.	CRF Expt. 1	Expt. 2	Difference	Mean	Estimated variance	Degrees of freedom
1	9.8	10.2	−0.4	10.0	0.08	1
2	9.3	9.7	−0.4	9.5	0.08	1
3	10.8	11.2	−0.4	11.0	0.08	1
4	10.6	10.8	−0.2	10.7	0.02	1
5	9.1	9.5	−0.4	9.3	0.08	1
6	9.5	10.1	−0.6	8.8	0.18	1
7	11.9	11.9	0.0	11.9	0.00	1
8	11.6	11.8	−0.2	11.7	0.02	1
				SUM	0.54	8
				Pooled	0.0675	

Table 10 *Recalculated full ANOVA table using the pooled run variance as the estimate of the residual variance*

Source of variation	SSQ	DF	MS	F value
Acetic acid concentration (A)	0.28125	1	0.28125	4.17
Methanol concentration (M)	7.41125	1	7.41125	109.80
Citric acid concentration (C)	0.03125	1	0.03125	0.46
Interactions AM	0.03125	1	0.03125	0.46
AC	0.00125	1	0.00125	0.02
MC	1.36125	1	1.36125	20.17
AMC	0.00125	1	0.00125	0.02
Total	9.11875	7		

MC, although much smaller than M, is significant at 95% confidence and warrants further investigation.

Interactions are very important in establishing robust analytical methodology. Many analytical chemists were taught at school, and alas in some instances at university, that THE way to conduct experiments was to vary one variable at a time and hold all others constant. This is probably the cardinal sin of experimentation. Analytical chemistry abounds with examples of where the level of one reagent non-linearly affects the effect of another. An easy way to look at this is to plot the CRF values observed for one factor at each of its levels for two levels of another.

For example, if the acetic acid concentration is plotted against the mean CRF for the two methanol levels the picture in Figure 14 is obtained.

Note that the lines are almost parallel indicating that there is no significant interaction. This is confirmed by the lack of significance for AM in Tables 7 and 10.

If the exercise is repeated for the methanol–citric acid concentration interaction, MC, the plot in Figure 15 results. Here the lines are clearly non-parallel and support the view that this interaction may well be analytically significant.

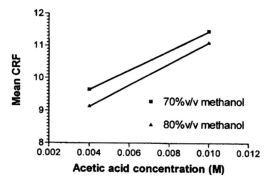

Figure 14 *Interaction plot for AM, acetic acid–methanol*

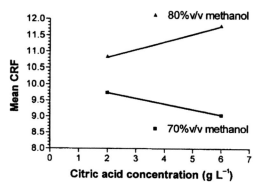

Figure 15 *Interaction plot for MC, methanol–citric acid*

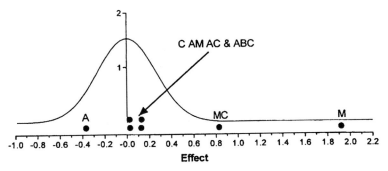

Figure 16 *Normal distribution for a residual variance of 0.0675 with effects plotted*

Pictures are usually more helpful than mere numbers in deciding whether a factor is important or not. Using the data in Table 10 and calculating what the normal distribution for a mean of 0 and a variance of 0.0675 would look like using a plotting programme is illustrated in Figure 16. The effects data are plotted along the x-axis. The MC value appears just beyond the tail of the residual error distribution and is certainly worth investigating further.

6.3 Multifactor experimental designs

For many method development problems, three or four factors are often the norm. The message is clearly that a simple approach to experimental design can be a crucial tool in ascertaining those factors which need to be controlled in order to maximise method robustness. In this example, the level of citric acid will have to be tightly controlled, as well as the methanol concentration, if consistent and high values of CRF are to be regularly obtained.

Three-level fractional factorial designs are also very useful, and charting the effects can be very helpful especially where there are more than three factors. The Plackett–Burman designs are often used to confirm (or otherwise!) the robustness of a method from the set value. Figure 17 shows some results[32] from a ruggedness study for an HPLC method for salbutamol[27] where the resolution factor, R_s, between it and its main degradation product is critical.

Note how in this instance the column-to-column variability is so large that the suitability for use must certainly be questioned.

Optimisation methods may also be used to maximise key parameters, *e.g.* resolution, but are beyond the scope of this handbook. Miller and Miller's book on *Statistics for Analytical Chemistry*[33] provides a gentle introduction to the topic of optimisation methods and response surfaces as well as digestible background reading for most of the statistical topics covered in this handbook. For those wishing to delve deeply into the subject of chemometric methods, the *Handbook of Chemometrics and Qualimetrics*[34] in two volumes by Massart *et al.*, is a detailed source of information.

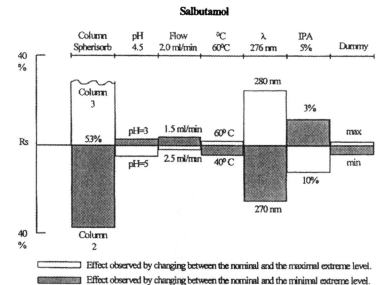

Salbutamol

Effect observed by changing between the nominal and the maximal extreme level.

Effect observed by changing between the nominal and the minimal extreme level.

Figure 17 *The effects of the different factors on the resolution factor, R_s, for salbutamol and its major degradation product*
(Reprinted from *Chromatographia*, 1998, **25**, 769. © (1998) Vieweg-Publishing)

7 Method Validation

The overall process of method validation is illustrated in Figure 1. However, the extent and scope of validation is governed by the applicability of the method.[23] An in-house procedure requires a less exacting process than a method intended for multi-matrix and/or multi-laboratory use. For the latter methods, a full collaborative trial is necessary and is covered in Chapter 9. However, for many purposes validation is limited to either demonstrating that method performance criteria established during development are met under routine laboratory conditions and/or showing method equivalence (Figure 18).

7.1 Recommended best practice for method validation

The intention of this section is to provide a framework for validation, not a comprehensive set of requirements. It should be regarded as a minimum. The implementation of a validation exercise should be customised for each application and the documented intent contained in a validation or verification protocol as outlined in NMLK No. 4.[35]

The United States Pharmacopoeia[36] identifies three categories of assay.

I   Analytical methods for quantitation of major components of bulk drug substances or active ingredients (including preservatives) in finished pharmaceutical products.

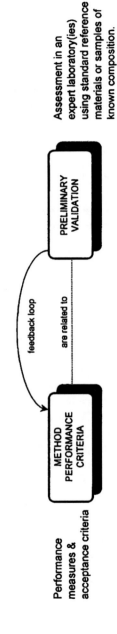

Figure 18 *Preliminary validation*

Table 11 USP24 *method validation guidelines*

	Assay category			
		II		
Analytical performance parameter	I	Quantitative	Limit tests	III
Accuracy	Yes	Yes	Possibly	Possibly
Precision	Yes	Yes	No	Yes
Specificity	Yes	Yes	Yes	Possibly
Limit of detection	No	No	Yes	Possibly
Limit of quantitation	No	Yes	No	Possibly
Linearity	Yes	Yes	No	Possibly
Range	Yes	Yes	Possibly	Possibly
Ruggedness	Yes	Yes	Yes	Yes

II Analytical methods for determination of impurities in bulk drug substances or degradation products in finished pharmaceutical products.

III Analytical methods for determination of performance characteristics (*e.g.* dissolution, drug release).

Guidelines for the selection of analytical performance parameters required for method validation are given in Table 11.

There is a remarkable degree of harmonisation between the USP approach[36] and the ICH 3 Note for Guidance on validation of analytical methods for the pharmaceutical industry[9] and the NMLK No. 4 guidelines for the food industry.[35] The requirements for the latter two are given in Table 12. For more detailed discussion of the pharmaceutical requirements see ref. 28.

One of the unfortunate choices of nomenclature is the use of 'specificity' where what is actually required is 'selectivity'. Few analytical techniques are specific for a given analyte but generally can be made sufficiently selective for the purpose. Alas the term seems to be firmly embedded!

The unit operations detailed in Table 12 are generally well described and characterised in the literature. Chapter 2 contains a listing of the majority of the terms and their definition. Linearity is discussed as a separate topic in Section 8.2.

7.2 Describing and writing analytical methods

Why is it that many published methods of analysis do not work when they are applied outside the developer's laboratory? Why do so many collaborative trials fail to produce consistent results? Some of the reasons may lie with inadequacy of method validation, sample preparation or lack of specification of key analytical parameters. However, even for well developed procedures, the failure

Table 12 *Framework for requirements for method validation*

Requirement	NMLK No. 4[35]	ICH3[9]
Validation or verification plan or protocol	✓	✓
Reference materials		✓
Specificity	✓	✓
Standard curve or linearity	✓	✓
Trueness or accuracy	✓	✓
Precision (repeatability, intermediate precision and reproducibility)	✓	✓
Concentration and measuring range	✓	✓
Limit of detection (LOD)	✓	✓
Limit of quantification (LOQ)	✓	✓
Robustness	✓	✓
Sensitivity	✓	✓
Evaluation of results	✓	✓
Documentation	✓	✓
Method performance monitoring (system suitability)	✓	✓
Change control	✓	

rate is high. Far too often, the cause is simply that the method is poorly written or insufficiently detailed.

The purpose of a written analytical procedure must be to convey the instructions necessary to allow a *competent* analytical chemist to reproduce the procedures and measurement processes faithfully, to apply the method to their sample and to state the results with appropriate confidence. I stress the word *competent*. Method transfer can only be accomplished between appropriately trained and qualified personnel. There is an unfortunate belief, in some quarters, that if procedures are written in an all-embracing and exhaustively detailed manner then professionally trained and qualified staff are not always necessary. This is a dangerous and misguided belief.

Anyone involved in writing analytical procedures and methods for the first time generally underestimates the difficulty of the task until faced with the results of an unsuccessful transfer process. Why is it then that we have a dearth of guidelines for such a task? The major texts on analytical chemistry and analytical science do not contain such advice. Even recent books on the validation of analytical methods,[28] The Approved Text to the FECS Curriculum of Analytical Chemistry[37] and *Quality Assurance in Analytical Chemistry*,[38] excellent though they are in other areas, make cursory reference, if any, to the requirements for good detailed written procedures.

The Analytical Methods Committee's own compilation of official and standardised methods of analysis[39] is widely used and respected within the analytical community. The importance of standardised formats for method documentation has been emphasised by the AMC's guidelines for achieving quality in trace analysis.[40] They list 17 headings for inclusion in the documentation:

1 Scope and significance	10 Quality control regime
2 Summary of method	11 Sample preparation
3 Definitions	12 Procedure
4 Safety precautions	13 Calculations
5 Precision and bias	14 Reporting
6 Sources of error	15 Bibliography and references
7 Apparatus	16 Appendices
8 Reagents and materials	17 Other information
9 Calibration and standardisation	

Taking both these approaches into consideration, Table 13 gives a simplified framework within which to define an analytical procedure.

Table 13 *Structure for analytical documentation*

1 Scope and applicability
   - Samples
   - Analytes
   - Ranges
2 Description and principle of the method
3 Equipment
   - Specification
   - Calibration and qualification
   - Range of operability
4 Reference materials and reagents
   - Specification
   - Preparation
   - Storage
5 Health and safety
6 Sampling
   - Methods
   - Storage
   - Limitations
7 Analytical procedure
   - Preparation of samples
   - Preparation of standards
   - Critical factors
   - Detailed description of all steps
   - Typical outputs; chromatograms, spectra, *etc.*
8 Recording and reporting of data
   - Method
   - Rounding and significant figures
   - Data treatments
9 Calculation of results
   - Calibration model
   - Calculation methods
   - Assumptions and limitations
10 Method performance
   - Statistical measures
   - Control charting
11 References and bibliography

8 Data Evaluation, Transformation and Reporting

Modern analytical systems produce data at an alarming rate and in copious quantities. Hopefully, these data are generated from qualified systems using validated procedures. The task of distilling out information and thus gaining knowledge can be a daunting and not inconsiderable task. The 'data to information' iceberg model, previously discussed in Figure 6 is shown in a simplified form in Figure 19.

This distillation process is usually carried out using statistical procedures. However, as Sir Ronald Fisher remarked, for this process to be successful, the data set has to contain the information sought.

'The statistician is no longer an alchemist expected to produce gold from any worthless material. He is more like a chemist capable of assaying exactly how much value it contains, and capable also of extracting this amount, and no more.'

No amount of elegant statistics or chemometric procedures will rescue inadequate or unreliable data.

From an analytical viewpoint, statistical approaches can be subdivided into two types: Exploratory Data Analysis (EDA) and Confirmatory Data Analysis (CDA). Exploratory data analysis is concerned with pictorial methods for visualising data shape and for looking for patterns in multivariate data. It should always be used as a precursor for selection of appropriate statistical tools to confirm or quantify, which is the province of confirmatory data analysis. CDA is about applying specific tools to a problem, quantifying underlying effects and data modelling. This is the more familiar area of statistics to the analytical community.

Some basic aspects of EDA will be explored in Section 8.1, and in Section 8.2 the most frequently used CDA technique, linear regression analysis for calibration, will be covered. It is not intended to provide a statistical recipe approach to be slavishly followed. The examples used and references quoted are intended to guide rather than to prescribe.

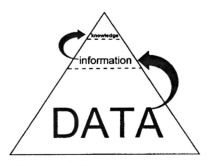

Figure 19 *The 'data to information' iceberg*

8.1 Exploratory data analysis

Exploratory data analysis, EDA, is an essential prerequisite of the examination of data by confirmatory methods. Time spent here can lead to a much greater appreciation of its structure and the selection of the most appropriate confirmatory technique. This has parallels in the analytical world. The story of the student's reply to the question 'Is the organic material a carboxylic acid?' which was 'I don't know because the IR scan isn't back yet' poses questions about the approaches to preliminary testing!

These EDA methods are essentially pictorial and can often be carried out using simple pencil and paper methods. Picturing data and displaying it accurately is an aspect of data analysis which is under utilised. Unless exploratory data analysis uncovers features and structures within the data set there is likely to be nothing for confirmatory data analysis to consider! One of the champions of EDA, the American statistician John W. Tukey, in his seminal work on EDA[41] captures the underlying principle in his comment that

'the greatest value of a picture is when it forces us to notice what we never expected to see'.

Take, for example, Anscombe's Quartet data set shown in Figure 20.[42] Here, there are four sets of 11 pairs of XY data which when subjected to the standard linear regression approach all yield identical numerical outputs. Cursory inspection of the numerical listing does not reveal the information immediately apparent when simply plotting the XY data (Figure 21). The Chinese proverb that a picture is worth ten thousand words is amply illustrated.

Other basic plotting techniques such as the dot plot can be helpful in the investigation of potential outliers in collaborative studies. For example, Figure 22 shows the results from an AMC trial on vitamin B_2 in animal feeding stuffs.

The simple EDA approaches developed by Tukey have been greatly extended by Tufte who remarked that

'Graphics reveal data. Indeed graphics can be more precise and revealing than conventional statistical calculations.'

I		II		III		IV	
X	Y	X	Y	X	Y	X	Y
10.0	8.04	10.0	9.14	10.0	7.46	8.0	6.58
8.0	6.95	8.0	8.14	8.0	6.77	8.0	5.76
13.0	7.58	13.0	8.74	13.0	12.74	8.0	7.71
9.0	8.81	9.0	8.77	9.0	7.11	8.0	8.84
11.0	8.33	11.0	9.26	11.0	7.81	8.0	8.47
14.0	9.96	14.0	8.10	14.0	8.84	8.0	7.04
6.0	7.24	6.0	6.13	6.0	6.08	8.0	5.25
4.0	4.26	4.0	3.10	4.0	5.39	19.0	12.50
12.0	10.84	12.0	9.13	12.0	8.15	8.0	5.56
7.0	4.82	7.0	7.26	7.0	6.42	8.0	7.91
5.0	5.68	5.0	4.74	5.0	5.73	8.0	6.89

$N = 11$
mean of X's = 9.0
mean of Y's = 7.5
equation of regression line: $Y = 3 + 0.5X$
standard error of estimate of slope = 0.118
$t = 4.24$
sum of squares $X - \bar{X} = 110.0$
regression sum of squares = 27.50
residual sum of squares of Y = 13.75
correlation coefficient = .82
$r^2 = .67$

Figure 20 *Anscombe's Quartet data set*[43]

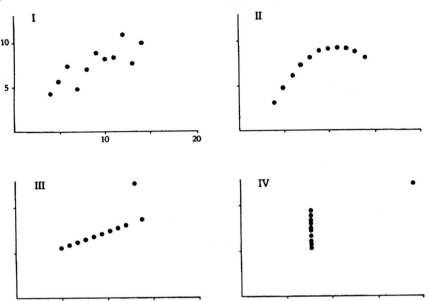

Figure 21 *Plotting Anscombe's Quartet* XY *data*[43]

Tufte's books on this topic, *The Visual Display of Quantitative Information* (1983),[44] *Envisioning Information* (1990)[45] and *Visual Explanations* (1997)[46] are recommended for further reading.

The availability of computer packages for data plotting and manipulation has made EDA easier to carry out. However, it is easy and instructive to carry out some simple EDA calculations by hand. Getting a feel for the shape of data doesn't always require the sophistication of a computer application: even reasonably large data sets can be handled using pencil and paper. The data in Table 14 are from an inter-laboratory study of a single sample for % protein and a single determination. They have been ranked, *i.e.* sorted, in ascending order.

One way of visualising the data set structure is to use the stem and leaf method. This simple display method defines the STEM as the number to the left of the decimal point. Note that as in Figure 23 for larger data sets it is usual to subdivide the STEM into two parts, *e.g.* values from 30.0 to 30.4 are recorded on the first row and 30.5 to 30.9 on the second. This split is sometimes indicated with a symbol, *e.g.* *, appended to the second row to make collation easier. The LEAF is the digit to the right of the decimal point for one decimal digit. Depending on the numbers of decimal places, an appropriate choice of scaling can be made. In this instance, the leaf unit is equivalent to 0.1% of protein. Tukey's book[41] is recommended for more details on this method and other EDA procedures.

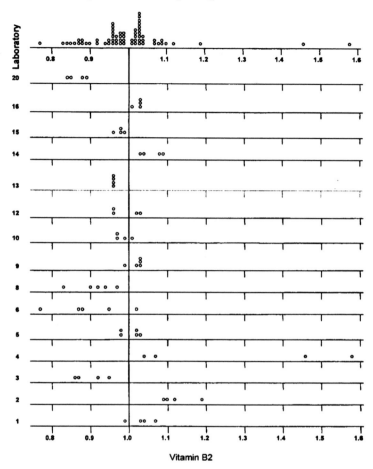

Figure 22 *Dot plot for normalised data from vitamin B₂ collaborative trial*

Stem	Leaves	Tally
27		0
27*	6	1
28		0
28*		0
29	022	3
29*	89	2
30		0
30*	677799	6
31	23333344444	11
31*	56666777777788888999999	23
32	0000011111111222222333344	25
32*	55666779	8
33	234	3
33*	55	2
	Overall sum	84

Figure 23 *Stem and leaf plot for % protein data*

Table 14 *% Protein data from 84 laboratories ranked in increasing order*

Laboratory	% Protein	Laboratory	% Protein	Laboratory	% Protein
83	27.6	32	31.7	31	32.2
20	29.0	33	31.7	40	32.2
47	29.2	48	31.7	44	32.2
75	29.2	49	31.7	51	32.2
84	29.8	59	31.7	56	32.2
63	29.9	6	31.8	4	32.3
16	30.6	9	31.8	29	32.3
2	30.7	13	31.8	34	32.3
38	30.7	58	31.8	65	32.3
55	30.7	74	31.8	7	32.4
67	30.9	26	31.9	24	32.4
69	30.9	28	31.9	1	32.5
80	31.2	30	31.9	53	32.5
27	31.3	52	31.9	18	32.6
39	31.3	72	31.9	36	32.6
41	31.3	77	31.9	42	32.6
62	31.3	5	32.0	35	32.7
71	31.3	19	32.0	79	32.7
8	31.4	37	32.0	25	32.9
11	31.4	66	32.0	46	33.2
15	31.4	68	32.0	21	33.3
45	31.4	10	32.1	61	33.4
73	31.4	23	32.1	64	33.5
22	31.5	54	32.1	81	33.5
12	31.6	57	32.1	**Mean**	**31.74**
43	31.6	60	32.1	**Median**	**31.9**
50	31.6	70	32.1	**Std dev.**	**0.97**
82	31.6	76	32.1	**RSD**	**3.05**
3	31.7	78	32.1		
14	31.7	17	32.2		

From Table 14, the basic stem and leaf plot can now be assembled. The resultant plot is shown in Figure 23.

The use of the tally column makes it easy to ensure that no data are missed. Ranking the data is not necessary but it makes it a lot easier to plot and to generate a box and whisker plot.

The shape of the distribution is quite apparent from Figure 23 and is much more information-rich than the ubiquitous histogram representation for a class interval of 0.5% protein. The overall shape of the data set is readily observed from the 'box and whisker' plot (Figure 24). This can be generated from the same ranked data set used for the stem and leaf plot. The principles are very straightforward and require only the ability to inspect the ranked data and perform a counting operation. For data ranked in increasing order the rules in Figure 25 apply.

For symmetrical data sets, the median lies centrally in the box and the whiskers are of equal length.

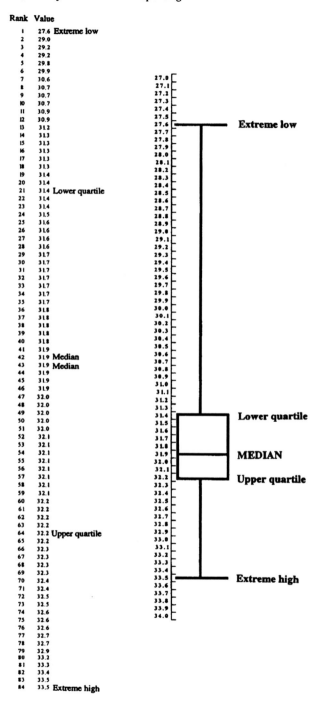

Figure 24 *Box and whisker plot of % protein data*

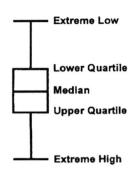

- Median value is the value that equally divides the data set; (data count +1)/2
- Inter-quartile distances are at the data count mid-points above and below the median
- The box contains 50% of the data set
- The whiskers go to the extreme values, or sometimes, to $1.5\times$ the inter-quartile distances. Values outside $\pm1.5\times$ the inter-quartile distances may be labelled as outliers

Figure 25 *Box and whisker plot principles*

Taking the data from Table 14, the simplest box and whisker plot using whiskers extending to the extreme values is shown in Figure 24. The median lies between rank 42 and 43 as the number of data points is even. In this instance the values are the same, 31.9%. The inter-quartile distances are at rank points 21 (31.4%) and 64 (32.2%) as they lie half way between the median and the lower and upper extreme values respectively. Note that here the possibility of there being outliers which should be excluded is ignored.

The construction of the plot requires only a pencil and ruler in this example to construct the essential features. The median is not central to the box and is skewed to the high side. However, the whiskers are not of equal length. The tail to the lower end of the distribution may indicate that some of the values may be outliers and should be investigated further.

It is hoped that the simple examples shown will encourage more EDA to be carried out before reaching for the statistical heavyweight procedures.

Many modern computer packages have routines that will generate stem and leaf and box and whisker plots as well as many more complicated ones for looking at multivariate data.

8.2 Linear calibration models

The most common calibration model or function in use in analytical laboratories assumes that the analytical response is a linear function of the analyte concentration. Most chromatographic and spectrophotometric methods use this approach. Indeed, many instruments and software packages have linear calibration (regression) functions built into them. The main type of calculation adopted is the method of least squares whereby the sums of the squares of the deviations from the predicted line are minimised. It is assumed that all the errors are contained in the response variable, Y, and the concentration variable, X, is error free. Commonly the models available are $Y = bX$ and $Y = bX + a$, where b is the slope of the calibration line and a is the intercept. These values are the least squares estimates of the true values. The following discussions are only

Table 15 *Spectrophotometric assay calibration data*

Data pair	Analyte (%w/v) X	Absorbance at 228.9 nm Y
1	0.000	0.001
2	0.013	0.058
3	0.020	0.118
4	0.048	0.280
5	0.100	0.579
6	0.123	0.735
7	0.155	0.866
8	0.170	1.007
9	0.203	1.149

illustrative of the least squares approach and standard works such as Draper and Smith[47] and Neter, Kutner, Nachtsheim and Wasserman[48] should be consulted for more details. For a gentler introduction, Miller and Miller[33] is recommended.

The first problem is deciding on which of these two common models to use. It has been argued that for spectrophotometric methods where the Beer–Lambert Law is known to hold, $Y = bX + \varepsilon$, the 'force through zero model' is the correct model to choose if the absorbance values are corrected for the blank.[49] The correct way to carry out the calibration regression is to include the blank response at 'assumed' zero concentration and use the model $Y = bX + a + \varepsilon$ instead. This may be a nicety from a practical standpoint for many assays but there are instances where a 'force through zero' model could produce erroneous results. Note that the ε denotes the random error term. Table 15 contains a set of absorbance concentration data from a UV assay.

The least squares regression approach will be illustrated using these data as an example for the model $Y = bX + a$. The calculations are shown in Table 16. The predicted least squares parameters are calculated by the following steps.

(1) Calculate the mean values for the X and Y data pairs. These are designated 'Xbar', $\bar{X}$, and 'Ybar', $\bar{Y}$, respectively.
(2) Calculate the values for the squares of $(X - \bar{X})$, $(Y - \bar{Y})$ and the cross product $(X - \bar{X})(Y - \bar{Y})$ for each data pair.
(3) Calculate from these the three sums of squares, SS_x, SS_y and SS_{xy}, respectively.
(4) The best estimate of the slope, b, is found by dividing SS_{xy} by SS_x.
(5) The best estimate of the intercept, a, is found from the equation $a = \bar{Y} - b\bar{X}$.

The resultant calibration plot is shown in Figure 26. The model equation is, $\hat{Y} = 5.7591X - 0.0002$ where $\hat{Y}$, 'Yhat', is the predicted Y value for a given X from best fit regression line. A plot of the residuals will give an indication of

Table 16 *Basic operations to calculate the least squares estimates of the model* $\hat{Y} = bX + a + \varepsilon$

Data pair	Analyte (%w/v) X	Absorbance at 228.9 nm Y	$(X - \bar{X})^2$	$(Y - \bar{Y})^2$	$(X - \bar{X})(Y - \bar{Y})$
1	0.000	0.001	0.0085	0.2826	0.0491
2	0.013	0.058	0.0063	0.2252	0.0377
3	0.020	0.118	0.0052	0.1719	0.0300
4	0.048	0.280	0.0020	0.0638	0.0112
5	0.100	0.579	0.0001	0.0022	0.0004
6	0.123	0.735	0.0009	0.0410	0.0062
7	0.155	0.866	0.0039	0.1112	0.0209
8	0.170	1.007	0.0060	0.2251	0.0368
9	0.203	1.149	0.0122	0.3800	0.0682
	0.092	0.533	0.0452	1.5028	0.2604
	$\bar{X}$	$\bar{Y}$	SS_x	SS_y	SS_{xy}

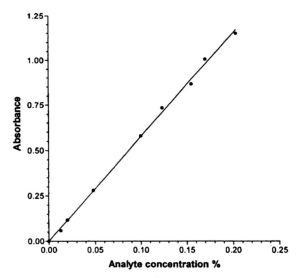

Figure 26 *Least squares calibration line for* $\hat{Y} = 5.7591X - 0.0002$

whether the model is well fitted. The residuals are listed in Table 17 and plotted in Figure 27.

There are no obvious features in the residuals plot to suggest that the model is unsuitable. However, some of the residuals are rather large so it is prudent to estimate how good our regression model is and what confidence we can have in data predicted from it. Estimates of the confidence in the determined slope and

Table 17 *Calculated residuals*

Data pair	Analyte (%w/v) X	Absorbance at 228.9 nm Y	$\hat{Y}$	Residual $(Y-\hat{Y})$
1	0.000	0.001	0.000	0.001
2	0.013	0.058	0.075	−0.017
3	0.020	0.118	0.115	0.003
4	0.048	0.280	0.277	0.003
5	0.100	0.579	0.576	0.003
6	0.123	0.735	0.709	0.026
7	0.155	0.866	0.893	−0.027
8	0.170	1.007	0.979	0.028
9	0.203	1.149	1.169	−0.020

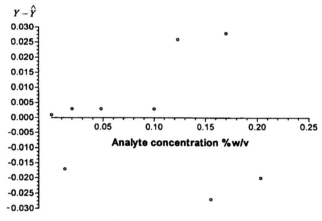

Figure 27 *Plot of residuals from* $\hat{Y} = bX + a + \varepsilon$

intercept parameters b and a and the overall regression can now be made. The steps are shown below and the results given in Table 18.

(1) The correlation coefficient, r, is calculated from

$$\sqrt{\frac{SS_{xy}^2}{SS_s.SS_y}} \qquad (7)$$

for all n data pairs.

(2) Calculate the mean square error, MSE, from

$$\frac{SS_y - bSS_{xy}}{n-2} \qquad (8)$$

where $n - 2$ are the degrees of freedom for the error.

Table 18 *Calculated regression and confidence parameters*

Correlation coefficient, r	0.9990
Mean square error, MSE	0.000417
Root mean square error, $RMSE$	0.0204
Standard error of the slope	0.0961
Standard error of the intercept	0.0112
Value of t for 7 degrees of freedom and 95% confidence	2.365
95% confidence limits of the slope	± 0.227
95% confidence limits of the intercept	± 0.0265

(3) Calculate the root mean square error, RMSE.
(4) Calculate the standard error of the slope from

$$\sqrt{\frac{MSE}{SS_x}} \tag{9}$$

(5) Calculate the standard error of the intercept from

$$\sqrt{MSE\left(\frac{\bar{X}^2}{SS_x} + \frac{1}{n}\right)} \tag{10}$$

(6) Calculate the 95% confidence limits for each by multiplying the standard errors by the value of t for $n - 2$ degrees of freedom.

The 95% confidence contours for regression can now be plotted on the regression line as is shown in Figure 28.

It is usual to test the slope of the regression line to ensure that it is significant using the F ratio, but for most analytical purposes this is an academic exercise. Of more importance is whether the intercept is statistically indistinguishable from zero. In this instance, the 95% confidence interval for the intercept is from -0.0263 to $+0.0266$ which indicates that this is the case.

Up until now it is only the confidence of the regression that has been examined. The impact of this calibration on predicted analyte concentrations from values of absorbance originating from unknown samples needs to be calculated.

The equation is a little daunting at first glance. The 95% confidence interval for the true concentration of the analyte, X, for a single determination is given by

$$\pm t\left(\frac{RMSE}{b}\right)\sqrt{\left[1 + \frac{1}{n} + \frac{\dfrac{(Y - \bar{Y})^2}{b}}{SS_x}\right]} \tag{11}$$

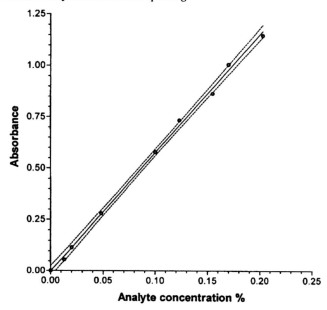

Figure 28 *Calibration line with 95% confidence limits for the regression*

If *m* replicates are carried out the equation becomes:

$$\pm t \left(\frac{RMSE}{b} \right) \sqrt{\left[\frac{1}{m} + \frac{1}{n} + \frac{(Y - \bar{Y})^2}{\dfrac{b}{SS_x}} \right]} \qquad (12)$$

For the calculation of simultaneous confidence intervals it is necessary to replace the *t* value in the above equations with $\sqrt{2F(2, n - 2)}$ from tables of the *F* distribution for the same degree of confidence. This is rarely done in practice but Figure 29 shows the superimposition of these limits on the calibration plot.

It is apparent from Figure 29 that the limits of prediction are considerably wider than those for regression and that the limits are much larger at the extremes of the regression than at the centre. This indicates that best practice is to devise calibrations sufficiently wide in range to allow the majority of samples to have values in the mid range if practicable.

A closer look at the prediction of the concentration for a single sample absorbance of 0.500 is shown in Figure 30. The least squares best fit value for the analyte concentration is 0.0868%. There is a 95% probability that the true concentration will lie ±0.0115% of the estimated value. The question is whether a 13% relative spread at 0.0868% is analytically acceptable. If this is not so, then the calibration procedure has to be reinvestigated to reduce variability.

When the *t* statistic version of the equations (11) and (12) is used, the arithmetic values for the 95% confidence limits are smaller. The corresponding

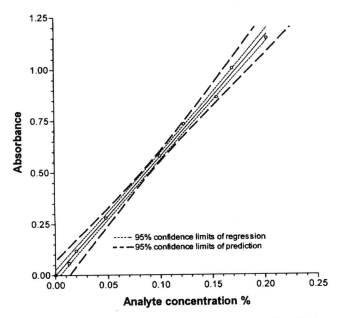

Figure 29 *Calibration plot with 95% limits of both regression and prediction*

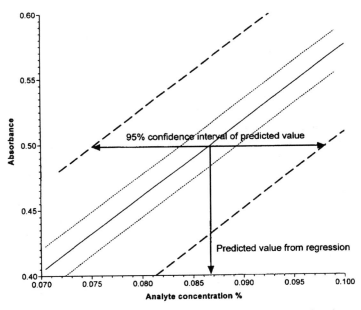

Figure 30 *Confidence in the predicted* X *concentration value for an absorbance of* 0.500 *from the calibration plot*

value for an analyte concentration of 0.0868% is ± 0.0089. The confidence limits can be tightened by making replicate measurements at each analyte concentration and using equation 12.

8.3 Recording and reporting of data

Good analytical practice has always demanded good record keeping. Proper and detailed records of all work carried out needs to be documented for future reference. One of the problems in carrying out this good practice is to decide how to record the numbers generated so that they reflect the correct number of significant figures and that these data are adequate for further calculations. There have been many examples where results have been recorded to eight decimal places because that was the number of digits displayed on the calculator! In this section, the topics of rounding of numbers will be considered along with the difficult issues of outliers in data sets and the expression of confidence in our results.

8.3.1 Rounding of numbers

The difference between the number of significant figures and the number of decimal places still causes problems. Take, for example, a result from a spectrophotometric assay of a material which is output from a calculator as 0.0119455. The number of significant figures is six whilst the number of decimal places is seven. The requirement is that the number of significant figures should reflect the experimental precision. Given the nature of spectrophotometry, the precision would normally affect the third significant figure or, in this example, the fourth decimal place. For statistical purposes it is usual to retain at least one more significant figure for calculations. The rounding process should be carried out only at the end of the calculation and in one step. The rule is that if the value of the least significant figure is 5 or more then round the preceding figure up. However, to prevent introducing an obvious bias when always rounding up with 5 a good practice is round to the nearest even number. Hence, 11.65 would round to 11.6 for three significant figures and 11.75 would round to 11.8.

As the error in the example is in the third significant figure, truncate the value at the fourth significant figure before rounding. The value to be rounded becomes 0.01194 so that rounding to three significant figures becomes 0.0119. Note that if we didn't truncate our data before rounding and started from the original six significant figures the results would be 0.011946 to five significant figures, 0.01195 to four significant figures and 0.0120 to three significant figures!

Further examples on the presentation of numerical values can be found in British Standard 1957: 1953 (confirmed 1987).

8.3.2 Outliers

The rejection of outliers from analytical data is one of the most vexed topics.[50] The only valid reason for rejecting a piece of data is when a discernible cause can be identified. Rejecting data on the grounds that it doesn't give the answer expected or appears anomalous is not acceptable. Equally unacceptable are the practices of slavish application of statistical outlier tests to make the excuse for the rejection. This does not mean that statistical outlier tests are useless. On the contrary, they are very helpful in directing attention to atypical results. It is the responsibility of the analytical practitioner to investigate atypical results, and statistical tools are useful in this regard but not prescriptive. The only exception is when using the IUPAC harmonised protocol for collaborative trials where this practice is accepted internationally. Even here, the requirement is that all data input from a trial have had to be screened for non-valid data.

The best advice is always to look carefully at the shape of the data set. Section 8.1 discussed various ways of exploratory data analysis. One method of spotting potential outliers to be investigated is to combine the z-score with a dot plot.

A z-score for any data point, X_i, in a data set of n values is calculated from

$$\frac{X_i - \bar{X}}{s_{n-1}} \tag{13}$$

where s_{n-1} is the standard deviation of the data set. Ideally, all the dots will be normally distributed about 0. A data set of 45 normalised vitamin B_2 data was transformed into z-scores and the resultant dot plot is shown in the upper trace of Figure 31. Most of the data are clustered between z-scores of ± 2 but two points are close to -3. If it is assumed that they are true outliers, then the s_{n-1} value can be recalculated excluding these points and the modified z-scores calculated. The lower trace in Figure 31 shows that most of the data remain clustered between z-scores of ± 2 but two points are beyond -4. These values are therefore worthy causes for a detailed investigation.

8.3.3 Confidence limits for analytical data

One of the key concerns of analytical science is 'how good are the numbers produced?'. Even with an adequately developed, optimised and collaboratively tested method which has been carried out on qualified and calibrated equipment the question remains. Recently it has become fashionable to extend the concepts of the physical metrology into analytical measurements and to quantify **confidence** in terms of the much more negative **uncertainty**.[54] It is based on the bottom-up principle or the so called error budget approach. This approach is based on the theory that if the variance contributions of all sources of error involved in analytical processes then it is possible to calculate the overall process

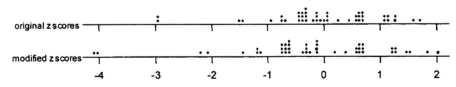

Figure 31 *Dot plots of z-scores for normalised vitamin B_2 data*

standard deviation and hence the uncertainty or confidence limits. Unfortunately, as Horwitz[52] has pointed out, this approach is likely to:

- overlook important variables and double count others;
- avoid considering unknown and unknowable interactions and interferences;
- adjust for missing variables with uncontrolled 'Type B' uncertainty components 'based on scientific judgement' estimated by dividing the tolerance by $\sqrt{3}$.

Physical metrology is typically dominated by systematic error with the random component being small in comparison. It is not unusual to have data reported to ten or more significant figures. Analytical data, on the other hand, are rarely able to be expressed to greater than four significant figures and for trace measurements only two or three. Modern analytical systems are too complex to make the use of the error budget viable particularly for trace analyses. Horwitz recommends the alternative 'top down approach' proposed by the AMC[53] for estimating uncertainty *via* collaborative trials.

9 Technology Transfer

When the method development process has been completed, an analytical procedure is subject to validation and transfer to routine use. This process may be called technology transfer which implies migration from the R&D environment to routine analytical laboratories. This process needs to be carried out whether the applicability of the procedure is limited to a single laboratory or to many laboratories.

9.1 Performance expectations and acceptance criteria

Analytical chemists have long been aware that the relative standard deviation increases as the analyte concentration decreases. Pioneering work by Horwitz[54] and the analysis of approximately 3000 values from collaborative trials has led to the establishment of an empirical function,

$$RSD = \pm 2^{(1-0.5\log C)} \tag{14}$$

which when plotted yields the Horwitz trumpet. This is illustrated in Figure 32.

This simple function is most useful in setting acceptance criteria for analytical method inter-laboratory precision. Note that the concentration, C, is relative so that 100% has a value of 1, *i.e.* 10^0. Table 19 lists values in decreasing powers of 10.

Other values of the Horwitz function can be calculated from the established analyte concentration value by substitution in $RSD = 2^{(1-0.5\log C)}$ remembering that C is the relative concentration.

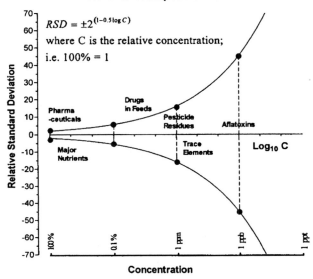

Figure 32 *Variation of relative standard deviation with concentration*
(Reprinted from *J. AOAC Int.*, 1980, **63**, 1334. © (1980) AOAC INTER-
NATIONAL)

Table 19 *Calculated values from the Horwitz function*

Relative concentration	RSD
10^0 (100%)	2.00
10^{-1} (10%)	2.83
10^{-2} (1%)	4.00
10^{-3} (0.1%)	5.66
10^{-4}	8.00
10^{-5}	11.31
10^{-6} (ppm)	16.00
10^{-7}	22.63
10^{-8}	32.00
10^{-9} (ppb)	45.25
10^{-10}	64.00
10^{-11}	90.51
10^{-12} (ppt)	128.00

When the value of RSD and the mean value of the analyte concentration have
been established, the Horwitz ratio, HORRAT, can be calculated.

$$HORRAT = \frac{RSD_{observed}}{RSD_{calculated}} \tag{15}$$

For example, if a collaborative trial produced a value for $RSD_{observed}$ of 9.21 for a mean analyte concentration of 0.00953% then

$$HORRAT = \frac{RSD_{obs}}{RSD_{calc}} = \frac{9.21}{8.06} = 1.14$$

It is now generally accepted practice that HORRAT values of 2 or less indicate that the method is of adequate precision. Hence, in our example, this is clearly so.

9.2 Transfer of published methods into a single laboratory

The Nordic Committee on Food Analysis has published a guideline on the 'Validation of Chemical Analytical Methods'[35] which differentiates between external validation work carried out on published methods and that work required to transfer it into the working laboratory to confirm its suitability for use. Table 20 is adapted from this document. This document is most easily accessed from a recently published book.[55] If certified reference materials are available they should be used to confirm the verification of trueness. These guidelines could also apply to intra-laboratory training programmes.

The overall validation requirements given in the guideline of the Nordic Committee on Food Analysis on the 'Validation of Chemical Analytical Methods'[35] are listed in Table 12.

Table 20 *Technology transfer validation requirements for published analytical methods*

Degree of external validation	Recommended internal validation
The method is externally validated in a method–performance study	Verification of trueness and precision
The method is externally validated in a method–performance study but is to be used with a new sample matrix or using new instruments	Verification of trueness and precision and where appropriate the detection limit
The method is well established but untested	More extensive validation/verification of key parameters
The method is published in the scientific literature and gives important performance characteristics	More extensive validation/verification of key parameters
The method is published in the scientific literature but without important performance characteristics	The method should be subject to full validation

9.3 Comparison of two methods

One of the most common tasks in the analytical laboratory is to determine whether or not two methods give comparable mean results within a predetermined confidence level. The test usually chosen for this task is the *t*-test.

There are two different ways of carrying out this test. The first one involves taking a single sample and analysing it by both methods a number of times. The usual procedure is to undertake a number of analyses (preferably not less than 6) for the chosen sample with both methods and calculate the value of the *t*-statistic. This is then compared with the tabular value for the appropriate degrees of freedom at the selected confidence level. If the calculated value is less than the tabulated *t* value then the mean values, and hence the methods, are accounted equivalent. This method has the advantage that the number of replicates undertaken for each method does not have to be equal. However, it is not always recognised that for this test to be valid the precision of the two methods should be equal. The method used to compare the precisions of methods is the *F*-ratio test and is carried out as part of the procedure.

The process may be illustrated with the following example. Table 21 gives some data from two methods, B and C, for estimating the silver content in photographic media. Method B is based upon an X-ray fluorescence procedure and Method C is a controlled potential coulometric method.

The steps and calculations are as follows.

(1) Calculate the differences and the square of the differences for each of the mean values of the data (n_1 data points from C and n_2 data points from B) from Methods C and B respectively.

Table 21 t-*Test comparison of means from two methods using a single sample*

Sample no.	Experimental result		Squares of deviations	
	Method C	Method B	Method C	Method B
1	0.8337	0.8741	0.000669	0.000010
2	0.8342	0.8733	0.000643	0.000006
3	0.8572	0.8715	0.000006	0.000000
4	0.8556	0.8738	0.000016	0.000008
5	0.8832	0.8485	0.000559	0.000501
6	0.8862	0.8513	0.000710	0.000384
7	0.8728	0.8483	0.000175	0.000512
8	0.8678	0.8478	0.000068	0.000535
9	0.8517	0.8904	0.000062	0.000380
10	0.8532	0.8862	0.000040	0.000234
11		0.8899		0.000360
12		0.8958		0.000619
Sum	8.5956	10.4507	0.002947	0.003550
	0.8596	0.8709	0.0003275	0.0003227
	Mean		**Variance**	

(2) Calculate the sums of these squares.

(3) Calculate the variances of Methods C and B (s_1^2 and s_2^2) by dividing the sum of squares by the degrees of freedom ($n_1 - 1$ and $n_2 - 1$ respectively).

(4) Calculate the F ratio by dividing the larger variance by the smaller. In this instance variance of C divided by variance of B.

(5) Compare this value with the tabular value (one-sided) for F at 95% confidence for 9 and 11 degrees of freedom.

(6) Calculate the pooled variance,

$$s^2 = \frac{\{(n_1 - 1)s_1^2 + (n_2 - 1)s_2^2\}}{(n_1 + n_2 - 2)} \tag{16}$$

(7) Calculate the absolute value for t from

$$|t| = \frac{(\bar{X}_1 - \bar{X}_2)}{s\sqrt{\left(\dfrac{1}{n_1} + \dfrac{1}{n_2}\right)}} \tag{17}$$

where t has $(n_1 + n_2 - 2)$ degrees of freedom.

(8) Compare this value with the tabulated values of t for $(n_1 + n_2 - 2)$ degrees of freedom (given in Table 22).

As the tabular t value exceeds the calculated value, then we can conclude that the means are not significantly different at 95% confidence.

One way of making this comparison visual is to generate pictures of distributions from the mean and variance data for Methods B and C and superimposing their data points. This assumes that the data are normally distributed but it has been shown that this is generally the case.[56] These are shown in Figure 33. The degree of overlap is high, and, importantly, the shapes of the distributions are very similar. This visual representation confirms the t statistic findings and supports the requirement for approximately equal variances.

However, there are occasions when variance equivalence is not observed. Another method, D, was used to determine the silver content of the photo-

Table 22 *Results from t-test comparison of means from two methods using a single sample*

F ratio of variances calculated	1.01
Tabular F (0.05,9,11) one-sided	2.91
Pooled variance, s^2	0.0003249
Standard deviation, s	0.018029
t Statistic	−1.47
Tabular t(0.05,20)	2.09

Comparison of Methods B and C

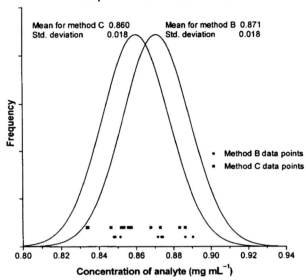

Figure 33 *Data distributions for Methods B and C and their data points*

graphic materials which was based on high precision automated emulsion densitometry. The equivalent picture to Figure 33 for this comparison is shown in Figure 34.

An F ratio test is hardly needed to decide whether the variances are different! A t-test is still applicable but needs to be modified to take into account the variance differences. This is done by calculating the effective number of degrees of freedom using Satterthwaite's method.[57] This is still an area which is controversial and a number of differing approaches and equations have been proposed.

Table 23 shows the data tabulations for Methods C and D ($n = 10$ and $n = 12$, respectively), analogous to Table 21 for Methods B and C. Note the large value for the F ratio indicating highly significant differences between the variances of the two data sets.

The steps and calculations for the unequal variance calculation are as follows.

(1) Calculate the sums and means for the first two columns.
(2) Calculate the squared differences between the mean for each method and the individual values and the sums for the second two columns.
(3) Calculate the variances for Methods C and D, V_C and V_D, by dividing the sums for columns 3 and 4 by (the number of data points − 1), in this example, 9 and 11.
(4) Calculate the F ratio by dividing the larger variance by the smaller, V_C/V_D.

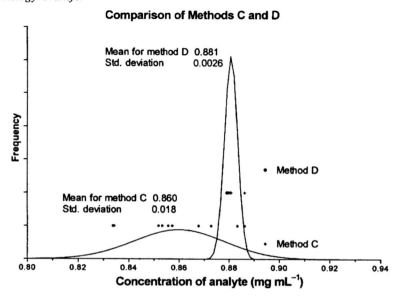

Figure 34 *Data distributions for Methods C and D and their data points*

Table 23 t-*Test calculations where the variances are known to be unequal (Satterthwaite's method)*

Sample no.	Experimental result		Squares of deviations	
	Method C	Method B	Method C	Method B
1	0.8337	0.8788	0.00066874	0.00000361
2	0.8342	0.8788	0.00064313	0.00000361
3	0.8572	0.8791	0.00000557	0.00000256
4	0.8556	0.8791	0.00001568	0.00000256
5	0.8832	0.8794	0.00055885	0.00000169
6	0.8862	0.8795	0.00070969	0.00000144
7	0.8728	0.8800	0.00017530	0.00000049
8	0.8678	0.8800	0.00006790	0.00000049
9	0.8517	0.8862	0.00006178	0.00003025
10	0.8532	0.8862	0.00004045	0.00003025
11		0.8807		0.00000000
12		0.8806		0.00000001
Sum	8.5956	10.5684	0.002947	0.00007696
	0.8596	0.8807	3.2745×10^{-4}	6.9964×10^{-6}
	Mean		**Variance**	

F ratio calculated	46.80
Tabular $F(0.05,9,11)$ one-sided	2.91

(5) Calculate the difference in the mean values of Methods C and D, $\bar{X}_C - \bar{X}_D$.

(6) Calculate the standard error between the method means from

$$\sqrt{(V_C/n_C + V_D/n_D)} \tag{18}$$

(7) Calculate the value for t from

$$\frac{(\bar{X}_C - \bar{X}_D)}{\sqrt{(V_C/n_C + V_D/n_D)}} \tag{19}$$

(8) Calculate the effective number of degrees of freedom from

$$\left\{ \frac{(V_C/n_C + V_D/n_D)^2}{\dfrac{(V_C/n_C)^2}{n_C - 1} + \dfrac{(V_D/n_D)^2}{n_D - 1}} \right\} \tag{20}$$

noting that this is rarely a whole number and the degrees of freedom result usually requires rounding. In this instance the value is 9.39.

(9) Look up the tabulated value of t for the effective number of degrees of freedom, 9, at 95% confidence.

(10) Compare this value with the calculated value of t found in step 7, as in Table 24.

As the tabular t value (2.26) is less than the absolute calculated value (the modulus) (3.66), then we can conclude that the means are significantly different at 95% confidence.

Up till now we have been concerned with comparing methods with a single sample. However, on many occasions this is not possible. For example, it may be that sample quantities do not permit such testing, the frequency of suitable sampling may be low so that testing has to take place over an extended period of time or a range of samples is desired to be examined by two methods. In such circumstances the paired version of the t-test is applicable. However, beware that although the requirements of variance equivalence are not needed, it is

Table 24 *Results from* t-test *calculations where the variances are known to be unequal*

Difference in means	−0.0211	
Standard error between means	0.0058	
t *Statistic*	−3.66	
Mean variances	3.2745×10^{-4}	6.9964×10^{-6}
Effective number of degrees of freedom	9.39	
t *Value of 9 effective degrees of freedom*	2.26	

assumed that for each method the variance is constant over the concentration range.

Consider the requirement for comparing two methods covering a number of sample types or matrices. Here the paired *t*-test is very useful. In the following example, eight different samples have been analysed by both methods with one determination each.

The steps and calculations are similar to those carried out previously.

(1) Calculate the mean for the differences between the data pairs, $\bar{X}_d$.
(2) Calculate the squared differences between the mean of the differences and the differences for each data pair.
(3) Calculate the variance of the differences by dividing the sum of squares and dividing it by (the number of data points − 1), in this example, 7.
(4) Calculate the standard deviation of the differences from the square root of the variance, s_d.
(5) Calculate the value for *t* from

$$\frac{\bar{X}_d \sqrt{n}}{s_d} \tag{21}$$

(6) Look up the tabulated value of *t* for the number of degrees of freedom $(n - 1)$, 7 in this instance, at 95% confidence (two-tailed test).
(7) Compare this value with the calculated value of *t* found in step 5.

The calculations are shown in Table 25. As the tabulated value is much greater than the calculated value the mean values derived from the two methods are not statistically significantly different.

Table 25 *Paired* t-*test calculation*

Sample pair	Method 1	Method 2	Difference	(Difference − mean difference)^2
1	0.551	0.540	0.011	0.000356
2	1.202	1.237	−0.035	0.000736
3	1.360	1.333	0.027	0.001216
4	1.694	1.657	0.037	0.002014
5	1.661	1.685	−0.024	0.000260
6	1.177	1.170	0.007	0.000221
7	1.627	1.680	−0.053	0.002036
8	1.201	1.234	−0.033	0.000631
Mean	1.309	1.317	−0.008	
Variance			0.001067	
Standard deviation			0.03267	
t calculated			−0.68	
t(0.05,7)			2.37	

9.4 Restricted inter-laboratory trials

After many years of debate and discussion, there is now an international consensus about the statistical approach and method to be employed in full collaborative trials.[21] This topic is discussed further in Section 9.5. The IUPAC protocol referred to above requires the analysis of duplicate test samples of the same material (a minimum of five materials or, exceptionally, three) in eight or more laboratories. However, there are many occasions where inter-laboratory studies are needed which, for various reasons, cannot achieve the prescribed criteria. In instances where the IUPAC criteria cannot be achieved, it is recommended that the Youden Matched Pairs procedure is used. The statistics of the method have been recently updated and a new procedure described.[58]

The procedure requires that there are two different but similar samples of the material to be analysed. Each sample is analysed once by each laboratory in the trial. The method is illustrated by using a data set of % aluminium in two limestone samples (X and Y) for ten laboratories taken from ref. 58 and shown in Table 26. The Youden plot[59] of these data is shown in Figure 35.

The graph is obtained by plotting X_i against Y_i results for each of the ten laboratories. The axes are drawn such that the point of intersection is at the mean values for X_i and Y_i. As a single method is used in the trial, the circle represents the standard deviation of the pooled X and Y data. The plot shows the predominance of systematic error over random error. Ideally, for bias-free data (*i.e.* containing no systematic error) the points would be clustered around the mid-point with approximately equal numbers in each of the four quadrants formed by the axes. In practice the points lie scattered around a 45° line. This pattern has been observed with many thousands of collaborative trials.

Table 26 *% Aluminium in limestone samples trial data*

Lab.	Test samples	
	X_i	Y_i
1	1.35	1.57
2	1.38	1.11
3	1.35	1.33
4	1.34	1.47
5	1.50	1.60
6	1.52	1.62
7	1.39	1.52
8	1.50	1.90
11	1.30	1.36
12	1.32	1.53
Mean	1.40	1.50

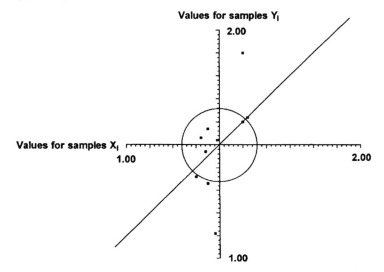

Figure 35 *Youden plot of % aluminium data for samples* X *and* Y

The calculations for the Youden matched pairs procedure are less complicated than those for the IUPAC protocol and do not involve outlier detection
and removal. For the example, the results are shown in Table 27.

Table 27 *Basic calculations for the Youden matched pairs procedure*

Lab.	Test samples					
	X_i	Y_i	Mean	$(Mean - Z)^2$	$(X_i - Z)^2$	$(Y_i - Z)^2$
1	1.35	1.57	1.460	0.0001	0.009604	0.014884
2	1.38	1.11	1.245	0.0412	0.004624	0.114244
3	1.35	1.33	1.340	0.0117	0.009604	0.013924
4	1.34	1.47	1.405	0.0018	0.011664	0.000484
5	1.50	1.60	1.550	0.0104	0.002704	0.023104
6	1.52	1.62	1.570	0.0149	0.005184	0.029584
7	1.39	1.52	1.455	0.0000	0.003364	0.005184
8	1.50	1.90	1.700	0.0635	0.002704	0.204304
11	1.30	1.36	1.330	0.0139	0.021904	0.007744
12	1.32	1.53	1.425	0.0005	0.016384	0.006724
Sum	13.95	15.01	14.48	0.1582	0.087740	0.420180
Mean	1.395	1.501	1.448			

Overall sample
mean, Z

No. of labs	10	
No. of samples	2	

The steps for the calculation procedure are as follows.

(1) Calculate the sums and means for the first two columns for samples X and Y.
(2) Calculate the laboratory mean for each pair (Column 3) and the overall mean for all samples, Z.
(3) Calculate the squares of the differences between the laboratory mean and the overall mean, Z, and their sum (Column 4).
(4) Calculate the squares of the differences between the individual values and the overall mean, Z, and their sum (Columns 5 and 6).

From these simple calculations the ANOVA table can be constructed together with the HORRAT ratio. An important difference from the conventional ANOVA table is that it allows for the sample variation to be accounted for. The general ANOVA table and calculated values for S samples are shown in Tables 28 and 29 but the example uses only two. However, if more than two samples are desired/available the procedure can be extended.

(5) Calculate the sums of squares and mean squares indicated in the ANOVA table.
(6) Calculate the repeatability from $S_r = \sqrt{MS_e}$.
(7) Calculate the reproducibility from

$$S_R = \sqrt{\frac{MS_L - MS_e}{2} + MS_e} \qquad (22)$$

(8) Calculate the observed RSD from

$$\left(\frac{S_R}{Z}\right) 100 \qquad (23)$$

Table 28 *General ANOVA table for Youden matched pairs for* S *samples and* L *laboratories*

Source	Degrees of freedom	Sum of squares	Mean squares
Laboratory (L)	$L - 1$	$SS_L = S\sum (Mean - Z)^2$	$MS_L = \dfrac{SS_L}{L-1}$
Sample (S)	$S - 1$	$SS_S = L\sum_S (\bar{X} - Z)^2 + (\bar{Y} - Z)^2 + \ldots$	$MS_S = \dfrac{SS_S}{S-1}$
Error (e)	$(L - 1)(S - 1)$	$SS_e = SS_T - SS_S - SS_L$	$MS_e = \dfrac{SS_e}{(L-1)(S-1)}$
Total	$LS - 1$	$SS_T = \sum_L (X_i - Z)^2 + (Y_i - Z)^2 + \ldots$	$MS_T = \dfrac{SS_T}{LS-1}$

Table 29 *Calculated values from the aluminium in limestone data set*

Source	Degrees of freedom	Value	Sum of squares	Mean squares
Laboratory (L)	$L - 1$	9	0.3163	0.03515
Sample (S)	1	1	0.0562	0.05618
Error (E)	$L - 1$	9	0.1354	0.01505
Total	$2L - 1$	19	0.5079	
Reproducibility	$S_R = \sqrt{\dfrac{MS_L - MS_e}{2} + MS_e} = 0.1584$			
Repeatability	$S_r = \sqrt{MS_e} = 0.1227$			
RSD	10.94			
Horwitz function	3.78			
HORRAT ratio	2.89			

(9) Calculate the Horwitz RSD from equation (14),

$$RSD = 2^{(1 - 0.5 \log C)}$$

where $C = Z/100$.

(10) Calculate the HORRAT ratio from equation (15),

$$RSD_{obs}/RSD_{calc}$$

It is readily apparent that the HORRAT ratio of 2.89 indicates that the method is not sufficiently precise for inter-laboratory work. However, inspection of the ANOVA table shows a large value for the mean square due to the samples, MS_S. This is an indication that they may not have been homogeneous and that the method may in fact be satisfactory. This demonstrates a major advantage in partitioning the variances between laboratory, sample and error rather than the traditional within- and between-laboratories method.

9.5 Full collaborative trials

As mentioned in the previous section, analytical methods intended for inter-laboratory use should be subjected to a collaborative trial using the IUPAC protocol. This is a prescriptive procedure and outlier testing/removal is mandatory. Before the process begins the data need to be examined to ensure that only valid data are input to the calculation process. This process is best shown as a flow chart (Figure 36).

The process looks very complicated but it is very structured and mechanical and is best shown by a worked example. The example chosen to demonstrate the

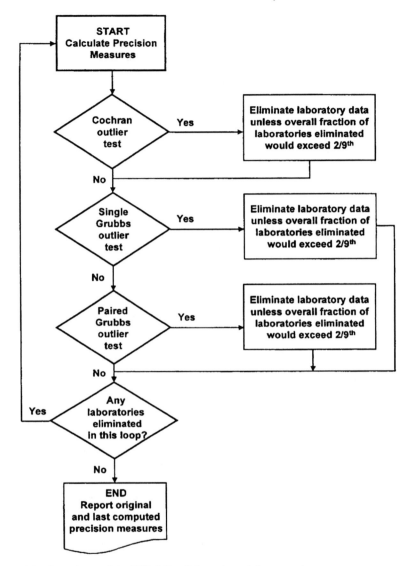

Figure 36 *Flow chart of the IUPAC collaborative trial protocol*
(Reprinted from *J. AOAC Int.*, 1995, **78**, 160A. © (1995) AOAC INTER-
NATIONAL)

calculation process is from an aflatoxin (ppb level) in peanut butter trial.[60]
These data are from 21 laboratories with one sample analysed in duplicate.
Clearly in a full IUPAC trial there would be three or more samples. The basic
data set and calculations are shown in Table 30.

For each laboratory, the within-laboratory variance is calculated from
$(R_1 - R_2)^2$ and the sum of the variances calculated. The first task is to identify

Table 30 *Aflatoxin in peanut butter data*

Lab.	Replicate 1	Replicate 2	All labs. Within-lab. variance	Lab. 21 removed Within-lab. variance	Labs. 21 and 5 removed Within-lab. variance
1	1.90	0.66	1.54	1.54	1.54
2	0.90	0.80	0.01	0.01	0.01
3	1.27	1.05	0.05	0.05	0.05
4	1.20	1.20	0.00	0.00	0.00
5	3.70	0.00	13.69	13.69	
6	0.90	1.20	0.09	0.09	0.09
7	0.00	0.00	0.00	0.00	0.00
8	1.70	1.60	0.01	0.01	0.01
9	0.00	1.20	1.44	1.44	1.44
10	0.60	0.90	0.09	0.09	0.09
11	1.30	0.90	0.16	0.16	0.16
12	0.70	1.70	1.00	1.00	1.00
13	1.40	1.40	0.00	0.00	0.00
14	1.00	0.20	0.64	0.64	0.64
15	0.80	1.10	0.09	0.09	0.09
16	0.00	0.00	0.00	0.00	0.00
17	2.10	0.60	2.25	2.25	2.25
18	0.00	0.00	0.00	0.00	0.00
19	0.90	1.10	0.04	0.04	0.04
20	1.60	1.50	0.01	0.01	0.01
21	7.20	12.50	28.09		
	Sum of the variances		49.20	21.11	7.42
	Cochran statistic (%)		57.10	64.86	30.34
	Critical value (%)		41.50	42.80	44.30

any single replicate which is too variable using Cochran's test. The Cochran statistic is the ratio of the highest laboratory variance to the total variance expressed as a percentage. In our example, Laboratory 21 is the highest with 28.09 which is 57.1% of the total. The critical value from Table A1 on Page 76 is 41.5 and therefore Laboratory 21's data are removed.

The next step is to examine the remaining data to see if a laboratory average or pair are too extreme in comparison with the rest of the data. This is somewhat more complicated and involves the calculation of Grubbs statistics but can be easily done with a spreadsheet. The process is as follows:

(1) calculate the standard deviation for the remaining 20 labs, s;
(2) calculate the standard deviation for the data excluding the highest average value, s_H;
(3) calculate the standard deviation for the data excluding the lowest average value, s_L;

(4) calculate the % decreases in standard deviation from

$$100\left(1 - \frac{s_H}{s}\right) \text{ and } 100\left(1 - \frac{s_L}{s}\right)$$

(5) the larger of these two numbers is the Grubbs single statistic;
(6) repeat the process again removing the second highest and second lowest values;
(7) the larger of these two numbers is the Grubbs pair statistic;
(8) look up the critical values from Column 1 in Table A2 on Page 77 for the Grubbs single statistic;
(9) look up the critical values from Column 2 in Table A2 for the Grubbs pair statistic.

Table 31 *First cycle Grubbs test calculations for aflatoxin data*

Lab.	Mean	Grubbs test (first cycle)			
1	1.28	1.28	1.28	1.28	1.28
2	0.85	0.85	0.85	0.85	0.85
3	1.16	1.16	0.16	1.16	1.16
4	1.20	1.20	1.20	1.20	1.20
5	1.85		1.85		1.85
6	1.05	1.05	1.05	1.05	1.05
7	0.00	0.00		0.00	
8	1.65	1.65	1.65	1.65	1.65
9	0.60	0.60	0.60	0.60	0.60
10	0.75	0.75	0.75	0.75	0.75
11	1.10	1.10	1.10	1.10	1.10
12	1.20	1.20	1.20	1.20	1.20
13	1.40	1.40	1.40	1.40	1.40
14	0.60	0.60	0.60	0.60	0.60
15	0.95	0.95	0.95	0.95	0.95
16	0.00	0.00	0.00	0.00	0.00
17	1.35	1.35	1.35		1.35
18	0.00	0.00	0.00	0.00	0.00
19	1.00	1.00	1.00	1.00	1.00
20	1.55	1.55	1.55	1.55	1.55
No. of labs	20	19	19	18	18
Overall mean	0.98	Highest removed	Lowest removed	2nd highest removed	2nd lowest removed
Calc. std. dev.	0.529 S	0.501 S_H	0.490 S_L	0.505 S_H	0.434 S_L
		Grubbs single statistic		Grubbs pair statistic	
		5.32%	7.46%	4.59%	18.01%
Grubbs test critical values		Single	23.6	Pair	33.2

The results of this process for our example are shown in Table 31. Neither of the calculated Grubbs statistics exceeds the critical value and therefore no laboratory is excluded.

It is now necessary to repeat the entire cycle with the remaining 20 laboratories. The next task is to recalculate the Cochran test statistic. In the example, Laboratory 5 has the highest remaining variance of 64.86, which is 57.1% of the new total variance of 21.11 (see Table 30).

The critical value from Table A1 is 42.8 and therefore Laboratory 5's data is removed. Note that this is allowed because the number of laboratories removed so far does not exceed 2/9th. In a similar manner, the Grubbs statistics are recalculated for this second cycle for the remaining 19 laboratories.

The results are shown in Table 32. Once again, neither of the critical values are exceeded so no new laboratories are excluded. As with the previous cycle the next task is to recalculate the Cochran test statistic. In the example, Laboratory

Table 32 *Second cycle Grubbs test calculations for aflatoxin data*

Lab.	Mean		Grubbs test (second cycle)		
1	1.28	1.28	1.28	1.28	1.28
2	0.85	0.85	0.85	0.85	0.85
3	1.16	1.16	1.16	1.16	1.16
4	1.20	1.20	1.20	1.20	1.20
6	1.05	1.05	1.05	1.05	1.05
7	0.00	0.00		0.00	
8	1.65	1.65	1.65		1.65
9	0.60	0.60	0.60	0.60	0.60
10	0.75	0.75	0.75	0.75	0.75
11	1.10	1.10	1.10	1.10	1.10
12	1.20	1.20	1.20	1.20	1.20
13	1.40	1.40	1.40	1.40	1.40
14	0.60	0.60	0.60	0.60	0.60
15	0.95	0.95	0.95	0.95	0.95
16	0.00	0.00	0.00	0.00	0.00
17	1.35	1.35	1.35		1.35
18	0.00	0.00	0.00	0.00	
19	1.00	1.00	1.00	1.00	1.00
20	1.55	1.55	1.55		1.55
No. of labs	19	18	18	17	17

		Highest removed	Lowest removed	2nd highest removed	2nd lowest removed
Overall mean	0.931				
Calc. std. dev.	0.501	0.484	0.461	0.469	0.402
	S	S_H	S_L	S_H	S_L

		Grubbs single statistic		Grubbs pair statistic	
		3.51%	8.10%	6.47%	19.82%
Grubbs test critical values		Single	24.6	Pair	34.5

Table 33 *Calculations for one way ANOVA for aflatoxin data after outlier removal*

Lab.	Replicate	X value	$X - \bar{X}$	$(X - \bar{X})^2$	Lab. mean	$X -$ Lab. mean	$(X -$ Lab. mean$)^2$	Lab. mean $- \bar{X}$	$($Lab. mean $- \bar{X})^2$
1	1	1.90	0.97	0.9389	1.28	0.62	0.3844	0.35	0.1218
1	2	0.66	-0.27	0.0735		-0.62	0.3844	0.35	0.1218
2	1	0.90	-0.03	0.0100	0.85	0.05	0.0025	-0.08	0.0066
2	2	0.80	-0.13	0.0172		-0.05	0.0025	-0.08	0.0066
3	1	1.27	0.34	0.1149	1.16	0.11	0.0121	0.23	0.0524
3	2	1.05	0.12	0.0141		-0.11	0.0121	0.23	0.0524
4	1	1.20	0.27	0.0723	1.20	0.00	0.0000	0.27	0.0723
4	2	1.20	0.27	0.0723		0.00	0.0000	0.27	0.0723
6	1	0.90	-0.03	0.0010	1.05	-0.15	0.0225	0.12	0.0141
6	2	1.20	0.27	0.0723		0.15	0.0225	0.12	0.0141
7	1	0.00	-0.93	0.8669	0.00	0.00	0.0000	-0.93	0.8669
7	2	0.00	-0.93	0.8669		0.00	0.0000	-0.93	0.8669
8	1	1.70	0.77	0.5913	1.65	0.05	0.0025	0.72	0.5169
8	2	1.60	0.67	0.4475		-0.05	0.0025	0.72	0.5169
9	1	0.00	-0.93	0.8669	0.60	-0.60	0.3600	-0.33	0.1096
9	2	1.20	0.27	0.0723		0.60	0.3600	-0.33	0.1096
10	1	0.60	-0.23	0.1096	0.75	-0.15	0.0225	-0.18	0.0328
10	2	0.90	-0.03	0.0010		0.15	0.0225	-0.18	0.0328
11	1	1.30	0.37	0.1361	1.10	0.20	0.0400	0.17	0.0285
11	2	0.90	-0.03	0.0010		-0.20	0.0400	0.17	0.0285
12	1	0.70	-0.23	0.0534	1.20	-0.50	0.2500	0.27	0.0723
12	2	1.70	0.77	0.5913		0.50	0.2500	0.27	0.0723
13	1	1.40	0.47	0.2199	1.40	0.00	0.0000	0.47	0.2199
13	2	1.40	0.47	0.2199		0.00	0.0000	0.47	0.2199
14	1	1.00	0.07	0.0048	0.60	0.40	0.1600	-0.33	0.1096
14	2	0.20	-0.73	0.5344		-0.40	0.1600	-0.33	0.1096
15	1	0.80	-0.13	0.0172	0.95	-0.15	0.0225	0.02	0.0004
15	2	1.10	0.17	0.0285		0.15	0.0225	0.02	0.0004
16	1	0.00	-0.93	0.8669	0.00	0.00	0.0000	-0.93	0.8669
16	2	0.00	-0.93	0.8669		0.00	0.0000	-0.93	0.8669
17	1	2.10	1.17	1.3664	1.35	0.75	0.5625	0.42	0.1755
17	2	0.60	-0.33	0.1096		-0.75	0.5625	0.42	0.1755
18	1	0.00	-0.93	0.8669	0.00	0.00	0.0000	-0.93	0.8669
18	2	0.00	-0.93	0.8669		0.00	0.0000	-0.93	0.8669
19	1	0.90	-0.03	0.0010	1.00	-0.10	0.0100	0.07	0.0048
19	2	1.10	0.17	0.0285		0.10	0.0100	0.07	0.0048
20	1	1.60	0.67	0.4475	1.55	0.05	0.0025	0.62	0.3831
20	2	1.50	0.57	0.3237		-0.05	0.0025	0.62	0.3831
Sum		35.38	0.00	12.7504			3.7080		9.0424
Mean		0.931		Total SS			Within-lab. SS		Between-lab. SS
No. of DPs		38							

Table 34 *ANOVA table and final calculated values for aflatoxin data*

Source	Degrees of freedom	Sum of squares	Mean squares
Within-laboratory	18	9.0424	0.5024
Between-laboratory	19	3.7080	0.1952
Total	37	12.7504	
Reproducibility	$S_R = \sqrt{\dfrac{MS_{Between} - MS_{Within}}{2} + MS_{Within}}$		0.5906
Repeatability	$S_r = \sqrt{MS_{within}}$		0.4418
RSD	63.64		
Horwitz value	45.25		
HORRAT ratio	1.40		

17 has the highest remaining variance of 2.25 which is 30.3% of the new total variance of 7.43 (see Table 30). The critical value from Table A1 is 44.3 for 19 laboratories and 2 replicates, and therefore Laboratory 17's data are not removed. As no more changes have taken place the third cycle for the Grubbs testing does not need to be done and the process is complete.

All of the foregoing calculations are simply to trim the data set by removing outliers in accordance with the set criteria. All that remains is to calculate the mean value for the sample, the repeatability and reproducibility of the trialled method and the HORRAT ratio. These values can be calculated using a one way ANOVA method similar to that described in Table 29 and given in Table 33 and the final values in Table 34.

The HORRAT ratio of 1.4 confirms that the method is sufficiently precise for use.

Postscript

An Excel spreadsheet, Handbook tables.xls, containing all of the calculation examples in this handbook, is to be made available for download, for teaching and information purposes only, from the Analytical Methods Committee home page. This can be accesed via the Books home page on the Royal Society of Chemistry web site http://www.rsc.org/is/books/vamp.htm.

The author and the Royal Society of Chemistry do not assume any liability in connection with its use.

Appendix: Statistical Tables

Table A1 *Critical values of Cochran's maximum variance ratio*[61]

No. of laboratories	r replicates per laboratory				
	r = 2	r = 3	r = 4	r = 5	r = 6
4	94.3	81.0	72.5	65.4	62.5
5	88.6	72.6	64.6	58.1	53.9
6	83.2	65.8	58.3	52.2	47.3
7	78.2	60.2	52.2	47.3	42.3
8	73.6	55.6	47.4	43.0	38.5
9	69.3	51.8	43.3	39.3	35.3
10	65.5	48.6	39.3	36.2	32.6
11	62.2	45.8	37.2	33.6	30.3
12	59.2	43.1	35.0	31.3	28.3
13	56.4	40.5	33.2	29.2	26.5
14	53.8	38.3	31.5	27.3	25.0
15	51.5	36.4	29.9	25.7	23.7
16	49.5	34.7	28.4	24.4	22.0
17	47.8	33.2	27.1	23.3	21.2
18	46.0	31.8	25.9	22.5	20.4
19	44.3	30.5	24.8	21.5	19.5
20	42.8	29.3	23.8	20.7	18.7
21	41.5	28.2	22.9	19.9	18.0
22	40.3	27.2	22.0	19.2	17.3
23	39.1	26.3	21.2	18.5	16.6
24	37.9	25.5	20.5	17.8	16.0
25	36.7	24.8	19.9	17.2	15.5
26	35.5	24.1	19.3	16.6	15.0
27	34.5	23.4	18.7	16.1	14.5
28	33.7	22.7	18.1	15.7	14.1
29	33.1	22.1	17.5	15.3	13.7
30	32.5	21.6	16.9	14.9	13.3
35	29.3	19.5	15.3	12.9	11.6
40	26.0	17.0	13.5	11.6	10.2
50	21.6	14.3	11.4	9.7	8.6

Reprinted from *J. AOAC Int.*, 1995, **78**, 158A. © (1995) AOAC INTERNATIONAL.

Table A2 *Critical values for Grubbs extreme deviation outlier tests*[62]

No. of laboratories	One highest or lowest	Two highest or two lowest
4	86.1	98.9
5	73.5	90.3
6	64.0	81.3
7	57.0	73.1
8	51.4	66.5
9	46.8	61.0
10	42.8	56.4
11	39.3	52.5
12	36.3[a]	49.1[a]
13	33.8	46.1
14	31.7	43.5
15	29.9	41.2
16	28.3	39.2
17	26.9	37.4
18	25.7	35.9
19	24.6	34.5
20	23.6	33.2
21	22.7	31.9
22	21.9	30.7
23	21.1	29.7
24	20.5	28.8
25	19.8	28.0
26	19.1	27.1
27	18.4	26.2
28	17.8	25.4
29	17.4	24.7
30	17.1	24.1
40	13.3	19.9
50	11.1	16.2

[a] The values for '12 laboratories' are missing from the original tables.[62] These values are obtained from the other tabulated data *via* cubic spline interpolation.
Reprinted from *J. AOAC Int.*, 1995, **78**, 159A. © (1995) AOAC INTERNATIONAL.

10 Selected publications of the AMC

C.A. Watson, ed., *Official and Standardised Methods of Analysis*, Royal Society of Chemistry, Cambridge, 3rd edn., 1994.

10.1 Statistics Sub-committee

- 'Recommendations for the Definition, Estimation and Use of the Detection Limit', *Analyst (London)*, 1987, **112**, 199.
- 'Recommendations for the Conduct and Interpretation of Cooperative Trials' *Analyst (London)*, 1987, **112**, 679.
- 'Uses (Proper and Improper) of Correlation Coefficients', *Analyst (London)*, 1988, **113**, 1469.
- 'Report on an Experimental Test of "Recommendation for the Conduct and Interpretation of Cooperative Trials"', *Analyst (Cambridge)*, 1989, **114**, 1489.
- 'Principles of Data Quality Control in Chemical Analysis', *Analyst (Cambridge)*, 1989, **114**, 1497.
- 'Robust Statistics – How not to Reject Outliers – Part 1 Basic Concepts', *Analyst (Cambridge)*, 1989, **114**, 1693.
- 'Robust statistics – How not to Reject Outliers – Part 2 Inter-laboratory Trials', *Analyst (Cambridge)*, 1989, **114**, 1699.
- 'Proficiency Testing of Analytical Laboratories: Organisational and Statistical Assessment', *Analyst (Cambridge)*, 1992, **117**, 97.
- 'Is my Calibration Linear?', *Analyst (Cambridge)*, 1994, **119**, 2363.
- 'Internal Quality Control of Analytical Data', *Analyst (Cambridge)*, 1995, **120**, 29.
- 'Uncertainty of Measurement: Implications for its Use in Analytical Science', *Analyst (Cambridge)*, 1995, **120**, 2303.
- 'Handling False Negatives, False Positives and Reporting Limits in Analytical Proficiency Tests', *Analyst (Cambridge)*, 1997, **122**, 495.
- 'Measurement Near Zero Concentration: Recording and Reporting Results that Fall Close To or Below the Detection Limit', in preparation.
- 'Bayesian Statistics – Can it Help Analytical Chemists?', in preparation.

10.2 Instrumental Criteria Sub-committee

Papers on the evaluation of analytical instrumentation:

- Part I   'Atomic Absorption Spectrophotometers, Primarily for use with Flames', *Anal. Proc.*, 1984, **21**, 45. Revised in *Analyst (Cambridge)*, 1998, **123**, 1407.
- Part II   'Atomic Absorption Spectrophotometers, Primarily for use with Electrothermal Atomizers', *Anal. Proc.*, 1985, **22**, 128. Revised in *Analyst (Cambridge)*, 1998, **123**, 1415.

- Part III   'Polychromators for use in Emission Spectrometry with ICP Sources', *Anal. Proc.*, 1986, **23**, 109.
- Part IV   'Monochromators for use in Emission Spectrometry with ICP Sources', *Anal. Proc.*, 1987, **24**, 3.
- Part V   'Inductively Coupled Plasma Sources for use in Emission Spectrometry', *Anal. Proc.*, 1987, **24**, 266.
- Part VI   'Wavelength Dispersive X-ray Spectrometers', *Anal. Proc.*, 1990, **27**, 324.
- Part VII   'Energy Dispersive X-ray Spectrometers', *Anal. Proc.*, 1991, **28**, 312.
- Part VIII   'Instrumentation for Gas–Liquid Chromatography', *Anal. Proc.*, 1993, **30**, 296.
- Part IX   'Instrumentation for High-performance Liquid Chromatography', *Analyst (Cambridge)*, 1997, **122**, 387.
- Part X   'Instrumentation for Inductively Coupled Plasma Mass Spectrometry', *Analyst (Cambridge)*, 1997, **122**, 393.
- Part XI   'Instrumentation for Molecular Fluorescence Spectrometry', *Analyst (Cambridge)*, 1998, **123**, 1649.
- Part XII   'Instrumentation for Capillary Electrophoresis', *Analyst (Cambridge)*, 2000, **125**, 361.
- Part XIII   'Instrumentation for UV–Visible–NIR Spectrometry', *Analyst (Cambridge)*, 2000, **125**, 367.
- Part XIV   'Instrumentation for Fourier Transform Infrared Spectrometry', *Analyst (Cambridge)*, 2000, **125**, 375.
- Part XV   'Instrumentation for Gas Chromatography–Ion-trap Mass Spectrometry', in preparation.
- Part XVI   'NIR Instrumentation for Process Control', in preparation.
- Part XVII   'NMR Instrumentation', in preparation.
- Part XVIII   'Instrumentation for Inductively Coupled Plasma Emission Spectrometry', revision and update of Parts III, IV and V, in preparation.

References

1. (a) BS ISO 5725–1: 1994, 'Accuracy (trueness and precision) of measurements, methods and results. Part 1: General principles and definitions'. (b) BS ISO 5725–1: 1994, 'Accuracy (trueness and precision) of measurements, methods and results. Part 2: A basic method for the determination of repeatability and reproducibility of a standard measurement method'. (c) BS ISO 5725–3: 1994, 'Accuracy (trueness and precision) of measurements, methods and results. Part 3: Intermediate measures of the precision of a test method'. (d) BS ISO 5725–4: 1994, 'Accuracy (trueness and precision) of measurements, methods and results. Part 4: Basic methods for estimating the trueness of a test method'. (e) BS ISO 5725–6: 1994, 'Accuracy (trueness and precision) of measurements, methods and results. Part 6: Use in practise of accuracy values'. (f) IUPAC, 'Nomenclature, symbols, units and their usage in spectrochemical analysis II. Data interpretation', *Anal. Chem.*, 1976, **48**, 2294. (g) 'Protocol for the design, conduct and interpretation of collaborative

studies', *Pure Appl. Chem.*, 1988, **60**, 855. (h) RSC Analytical Methods Committee, 'Recommendations for the definition, estimation and use of the detection limit', *Analyst (London)*, 1987, **112**, 199. (i) R. W. Stephany. 'Quality assurance and control in the analysis of foodstuffs, residue analysis in particular', *Belgium J. Food Chem.*, 1989, **44**, 139. (j) ISO Guide 33: 1989, 'Use of certified reference materials'. (k) British Standard 1957: 1953 (confirmed 1987), 'The presentation of numerical values'. (l) IUPAC, Nomenclature for Sampling in Analytical Chemistry. *Pure Appl. Chem.*, 1990, **62**, 1193.

2. J. Incédy, T. Lengyel and A.M. Ure, *IUPAC Compendium of Analytical Nomenclature Definitive Rules 1997*, Blackwell Science, Oxford, 3rd edn., 1998.
3. N.T. Crosby and I. Patel, *General Principles of Good Sampling Practice*, Royal Society of Chemistry, Cambridge, 1995.
4. F.E. Prichard (co-ordinating author), *Quality in the Analytical Chemistry Laboratory*, John Wiley, London, 1995, Ch. 2.
5. W. Horwitz, 'Nomenclature for Sampling in Analytical Chemistry', *Pure Appl. Chem.*, 1990, **62**, 1193.
6. (a) British Standard 6000: 1972, 'Guide to the use of BS6001, Sampling procedures and tables for inspection by attributes', (ISO 2859 part 0). (b) British Standard 6001: 1989, Parts 1 to 3, 'Sampling procedures for inspection by attributes' (ISO 2859 parts 1 to 3). (c) British Standard 6002: 1979, 'Sampling procedures and charts for inspection by variables for percent defective' (ISO 3951).
7. Food & Drug Administration, Center for Drug Evaluation and Research, Draft Guidance for Industry, August 1999, ANDAs: Blend Uniformity Analysis.
8. M. Thompson, S.L.R. Ellison, A. Fajgelj, P. Willetts and R. Wood, *Pure Appl. Chem.*, 1999, **71**, 337.
9. ICH3, 1995, Note for guidance on validation of analytical procedures: methodology {CPMP/ICH/281/95}.
10. FDA Center for Drug Evaluation and Research, Reviewer Guidance, 'Validation of Chromatographic Methods', 1994.
11. US Pharmacopoeia 24, 2000, ⟨621⟩ chromatography, ⟨711⟩ dissolution testing, ⟨831⟩ refractive index, ⟨851⟩ spectrophotometry and light scattering, *etc.*
12. BP 1999, Appendix II; Infrared spectrophotometry, Ultraviolet and Visible spectrophotometry and Pharm. Eur., 3rd edn., 1997 V.6.19 Absorption Spectrophotometry, *etc.*
13. Australian Code of GMP for Therapeutic Goods, 'Medicinal Products; Appendix D, Guidelines for Laboratory Instrumentation', 1991.
14. NIS 45, Edition 2, May 1996, Accreditation for Chemical Laboratories (EURACHEM Guidance Document No. 1/WELAC Guidance Document No. WGD 2 (Edition 1).
15. C. Burgess, *Laboratory Automation and Information Management*, 1995, **31**, 35.
16. C. Burgess, D.G. Jones and R.D. McDowall, *Analyst (Cambridge)*, 1998, **123**, 1879.
17. W.B. Furman, T.P. Layloff and R.T. Tetzlaff, *J. AOAC Int.*, 1994, **77**, 1314.
18. CITAC Guide 1: 1995, 'International Guide to Quality in Analytical Chemistry, An Aid to Accreditation', Edition 1.
19. VAM Instrumentation Working Group, 'Guidance on the Equipment Qualification of Analytical Instruments'; High Performance Liquid Chromatography (HPLC), 1998.
20. D. Parriott, *LC-GC*, 1994, **12**, 132ff.
21. Collaborative Study Guidelines, *J. AOAC Int.*, 1995, **78**, 143A–160A.
22. G.T. Wernimont, *Use of statistics to develop and evaluate analytical methods*, AOAC, Arlington, VA, 1985.
23. EURACHEM, 'The Fitness for Purpose of Analytical Methods', Edition 1.0, LGC (Teddington) Limited, 1998.
24. EURACHEM/CITAC Guide CG2, 'Quality Assurance for Research and Development and Non-routine Analysis', Edition 1.0, LGC (Teddington) Limited, 1998.

25. E.B. Sandel, 'Errors in Chemical Analysis', in *Treatise of Analytical Chemistry*, I. M. Kolthoff and P. J. Elving, eds., Part 1, Vol. 1, John Wiley, New York, 1959.
26. E. Morgan, *Chemometrics: Experimental Design*, ACOL Series, John Wiley, Chichester, 1991.
27. M.M.W.B. Hendriks, J.H. De Boer and A.K. Smilde, eds., *Robustness of Analytical Chemical Methods and Pharmaceutical Technological Products*, Elsevier, Amsterdam, 1996.
28. C.M. Riley and T.W. Rosanske, eds., *Development and Validation of Analytical Methods*, Pergamon Elsevier Science, Oxford, 1996.
29. S.N. Deming and S.L. Morgan, *Experimental Design: a Chemometric Approach*, Elsevier, Amsterdam, 2nd edn., 1993.
30. D.C. Montgomery, *Design and Analysis of Experiments*, John Wiley & Sons, New York, 4th edn., 1997.
31. G.E.P. Box, W.G. Hunter and J.S. Hunter, *Statistics for Experimenters*, John Wiley, London, 1978.
32. M. Mulholland and J. Waterhouse, *Chromatographia*, 1988, **25**, 769.
33. J.C. Miller and J.N. Miller, *Statistics for Analytical Chemistry*, Ellis Horwood, Chichester, 3rd edn., 1993.
34. D.L. Massart, B.G.M. Vandeginste, L.M.C. Budyens, S. De Jong, P.J. Lewi and J. Smeyers-Verbeke, *Handbook of Chemometrics and Qualimetrics*, Parts A and B, Elsevier, Amsterdam, 1997/1998.
35. NMLK Procedure No. 4, 1996, 'Validation of Chemical Analytical Methods', February 1997.
36. US Pharmacopoeia, 24, 2000, ⟨1225⟩.
37. R. Kellner, J.-M. Mermet, M. Otto and H.M. Widmer, eds., *Analytical Chemistry*, Wiley-VCH, Weinheim, 1998.
38. W. Funk, V. Dammann and G. Donnevert, *Quality Assurance in Analytical Chemistry*, VCH, Weinheim, 1995.
39. C.A. Watson, ed., *Official and Standardised Methods of Analysis*, Royal Society of Chemistry, Cambridge, 3rd edn., 1994.
40. M. Sargent and G. MacKay, *Guidelines for Achieving Quality in Trace Analysis*, Royal Society of Chemistry, Cambridge, 1995.
41. J.W. Tukey, *Exploratory Data Analysis*, Addison Wesley, Reading, PA, 1977.
42. F.J. Anscombe, *Am. Statistician*, 1973, **27**, 17.
43. Graphic display from reference 44.
44. E.R. Tufte, *The Visual Display of Quantitative Information*, Graphic Press, Cheshire, CT, 1983.
45. E.R. Tufte, *Envisioning Information*, Graphics Press, Cheshire, CT, 1990.
46. E.R. Tufte, *Visual Explanations*, Graphics Press, Cheshire, CT, 1997.
47. N.R. Draper and H. Smith, *Applied Regression Analysis*, John Wiley & Sons, New York, 3rd edn., 1998.
48. J. Neter, M.H. Kutner, C.J. Nachtsheim and W. Wasserman, *Applied Linear Regression Models*, Irwin, Chicago, IL, 3rd edn., 1996.
49. R. Caulcutt and R. Boddy, *Statistics for Analytical Chemists*, Chapman and Hall, London, 1983.
50. V. Barnett and T. Lewis, *Outliers in Statistical Data*, John Wiley, London, 3rd edn., 1994.
51. EURACHEM, *Quantifying Uncertainty in Analytical Measurement*, LGC (Teddington) Limited, 1995.
52. W. Horwitz, *J. AOAC Int.*, 1998, **81**, 785.
53. Analytical Methods Committee, *Analyst (Cambridge)*, 1995, **120**, 2203.
54. W. Horwitz, *Anal. Chem.*, 1982, **54**, 67A.
55. R. Wood, A. Nilsson and H. Wallin, *Quality in the Food Analysis Laboratory*, Royal Society of Chemistry, Cambridge, 1998.
56. M. Thompson and R.J. Howarth, *Analyst (London)*, 1980, **105**, 1188.

57. F.W. Satterthwaite, *Biom. Bull.*, 1946, **2**, 110.
58. F.D. McClure, *J. AOAC Int.*, 1999, **82**, 375.
59. W.J. Youden and E.H. Steiner, *Statistical Manual of the AOAC*, AOAC International, Arlington, VA, 1975.
60. F.M. Worner, A.L. Patey and R. Wood, *J. Assoc. Public Analysts*, 1992, **28**, 1.
61. *J. AOAC Int.*, 1995, **78**, 158A.
62. *J. AOAC Int.*, 1995, **78**, 159A.

Subject Index

Note: Figures and Tables are indicated (in this index) by *italic page numbers*

Printed in the United Kingdom by
Lightning Source UK Ltd., Milton Keynes
142007UK00001BA/49/A

9 780854 044825